Food: A Very Short Introduction

VERY SHORT INTRODUCTIONS are for anyone wanting a stimulating and accessible way in to a new subject. They are written by experts, and have been published in more than 25 languages worldwide.

The series began in 1995, and now represents a wide variety of topics in history, philosophy, religion, science, and the humanities. The VSI Library now contains more than 300 volumes—a Very Short Introduction to everything from ancient Egypt and Indian philosophy to conceptual art and cosmology—and will continue to grow in a variety of disciplines.

Very Short Introductions available now:

ADVERTISING Winston Fletcher
AFRICAN HISTORY John Parker
    and Richard Rathbone
AGNOSTICISM Robin Le Poidevin
AMERICAN HISTORY Paul S. Boyer
AMERICAN IMMIGRATION
    David A. Gerber
AMERICAN POLITICAL PARTIES
    AND ELECTIONS L. Sandy Maisel
AMERICAN POLITICS Richard M. Valelly
THE AMERICAN PRESIDENCY
    Charles O. Jones
ANAESTHESIA Aidan O'Donnell
ANARCHISM Colin Ward
ANCIENT EGYPT Ian Shaw
ANCIENT GREECE Paul Cartledge
ANCIENT PHILOSOPHY Julia Annas
ANCIENT WARFARE
    Harry Sidebottom
ANGELS David Albert Jones
ANGLICANISM Mark Chapman
THE ANGLO-SAXON AGE John Blair
THE ANIMAL KINGDOM
    Peter Holland
ANIMAL RIGHTS David DeGrazia
THE ANTARCTIC Klaus Dodds
ANTISEMITISM Steven Beller
ANXIETY Daniel Freeman and
    Jason Freeman
THE APOCRYPHAL GOSPELS Paul Foster
ARCHAEOLOGY Paul Bahn
ARCHITECTURE Andrew Ballantyne
ARISTOCRACY William Doyle
ARISTOTLE Jonathan Barnes

ART HISTORY Dana Arnold
ART THEORY Cynthia Freeland
ATHEISM Julian Baggini
AUGUSTINE Henry Chadwick
AUSTRALIA Kenneth Morgan
AUTISM Uta Frith
THE AVANT GARDE David Cottington
THE AZTECS David Carrasco
BACTERIA Sebastian G. B. Amyes
BARTHES Jonathan Culler
THE BEATS David Sterritt
BEAUTY Roger Scruton
BESTSELLERS John Sutherland
THE BIBLE John Riches
BIBLICAL ARCHAEOLOGY Eric H. Cline
BIOGRAPHY Hermione Lee
THE BLUES Elijah Wald
THE BOOK OF MORMON Terryl Givens
BORDERS Alexander C. Diener and
    Joshua Hagen
THE BRAIN Michael O'Shea
THE BRITISH CONSTITUTION
    Martin Loughlin
THE BRITISH EMPIRE Ashley Jackson
BRITISH POLITICS Anthony Wright
BUDDHA Michael Carrithers
BUDDHISM Damien Keown
BUDDHIST ETHICS Damien Keown
CANCER Nicholas James
CAPITALISM James Fulcher
CATHOLICISM Gerald O'Collins
THE CELL Terence Allen and
    Graham Cowling
THE CELTS Barry Cunliffe

CHAOS Leonard Smith
CHILDREN'S LITERATURE
    Kimberley Reynolds
CHINESE LITERATURE Sabina Knight
CHOICE THEORY Michael Allingham
CHRISTIAN ART Beth Williamson
CHRISTIAN ETHICS D. Stephen Long
CHRISTIANITY Linda Woodhead
CITIZENSHIP Richard Bellamy
CIVIL ENGINEERING David Muir Wood
CLASSICAL MYTHOLOGY
    Helen Morales
CLASSICS Mary Beard and
    John Henderson
CLAUSEWITZ Michael Howard
CLIMATE Mark Maslin
THE COLD WAR Robert McMahon
COLONIAL AMERICA Alan Taylor
COLONIAL LATIN AMERICAN
    LITERATURE Rolena Adorno
COMEDY Matthew Bevis
COMMUNISM Leslie Holmes
THE COMPUTER Darrel Ince
THE CONQUISTADORS Matthew
    Restall and Felipe Fernández-Armesto
CONSCIENCE Paul Strohm
CONSCIOUSNESS Susan Blackmore
CONTEMPORARY ART Julian Stallabrass
CONTEMPORARY FICTION
    Robert Eaglestone
CONTINENTAL PHILOSOPHY
    Simon Critchley
COSMOLOGY Peter Coles
CRITICAL THEORY Stephen Eric Bronner
THE CRUSADES Christopher Tyerman
CRYPTOGRAPHY Fred Piper and
    Sean Murphy
THE CULTURAL
    REVOLUTION Richard Curt Kraus
DADA AND SURREALISM
    David Hopkins
DARWIN Jonathan Howard
THE DEAD SEA SCROLLS Timothy Lim
DEMOCRACY Bernard Crick
DERRIDA Simon Glendinning
DESCARTES Tom Sorell
DESERTS Nick Middleton
DESIGN John Heskett
DEVELOPMENTAL BIOLOGY
    Lewis Wolpert

THE DEVIL Darren Oldridge
DIASPORA Kevin Kenny
DICTIONARIES Lynda Mugglestone
DINOSAURS David Norman
DIPLOMACY Joseph M. Siracusa
DOCUMENTARY FILM
    Patricia Aufderheide
DREAMING J. Allan Hobson
DRUGS Leslie Iversen
DRUIDS Barry Cunliffe
EARLY MUSIC Thomas Forrest Kelly
THE EARTH Martin Redfern
ECONOMICS Partha Dasgupta
EDUCATION Gary Thomas
EGYPTIAN MYTH Geraldine Pinch
EIGHTEENTH-CENTURY
    BRITAIN Paul Langford
THE ELEMENTS Philip Ball
EMOTION Dylan Evans
EMPIRE Stephen Howe
ENGELS Terrell Carver
ENGINEERING David Blockley
ENGLISH LITERATURE Jonathan Bate
ENVIRONMENTAL ECONOMICS
    Stephen Smith
EPIDEMIOLOGY Rodolfo Saracci
ETHICS Simon Blackburn
THE EUROPEAN UNION John Pinder
    and Simon Usherwood
EVOLUTION Brian and Deborah
    Charlesworth
EXISTENTIALISM Thomas Flynn
FASCISM Kevin Passmore
FASHION Rebecca Arnold
FEMINISM Margaret Walters
FILM Michael Wood
FILM MUSIC Kathryn Kalinak
THE FIRST WORLD WAR Michael Howard
FOLK MUSIC Mark Slobin
FOOD John Krebs
FORENSIC PSYCHOLOGY David Canter
FORENSIC SCIENCE Jim Fraser
FOSSILS Keith Thomson
FOUCAULT Gary Gutting
FREE SPEECH Nigel Warburton
FREE WILL Thomas Pink
FRENCH LITERATURE John D. Lyons
THE FRENCH REVOLUTION
    William Doyle
FREUD Anthony Storr

FUNDAMENTALISM Malise Ruthven
GALAXIES John Gribbin
GALILEO Stillman Drake
GAME THEORY Ken Binmore
GANDHI Bhikhu Parekh
GENIUS Andrew Robinson
GEOGRAPHY John Matthews and
    David Herbert
GEOPOLITICS Klaus Dodds
GERMAN LITERATURE Nicholas Boyle
GERMAN PHILOSOPHY
    Andrew Bowie
GLOBAL CATASTROPHES Bill McGuire
GLOBAL ECONOMIC HISTORY
    Robert C. Allen
GLOBAL WARMING Mark Maslin
GLOBALIZATION Manfred Steger
THE GOTHIC Nick Groom
GOVERNANCE Mark Bevir
THE GREAT DEPRESSION AND THE
    NEW DEAL Eric Rauchway
HABERMAS James Gordon Finlayson
HAPPINESS Daniel M. Haybron
HEGEL Peter Singer
HEIDEGGER Michael Inwood
HERODOTUS Jennifer T. Roberts
HIEROGLYPHS Penelope Wilson
HINDUISM Kim Knott
HISTORY John H. Arnold
THE HISTORY OF ASTRONOMY
    Michael Hoskin
THE HISTORY OF LIFE Michael Benton
THE HISTORY OF MATHEMATICS
    Jacqueline Stedall
THE HISTORY OF MEDICINE
    William Bynum
THE HISTORY OF TIME
    Leofranc Holford-Strevens
HIV/AIDS Alan Whiteside
HOBBES Richard Tuck
HUMAN EVOLUTION Bernard Wood
HUMAN RIGHTS Andrew Clapham
HUMANISM Stephen Law
HUME A. J. Ayer
IDEOLOGY Michael Freeden
INDIAN PHILOSOPHY Sue Hamilton
INFORMATION Luciano Floridi
INNOVATION Mark Dodgson and
    David Gann
INTELLIGENCE Ian J. Deary

INTERNATIONAL MIGRATION
    Khalid Koser
INTERNATIONAL RELATIONS
    Paul Wilkinson
ISLAM Malise Ruthven
ISLAMIC HISTORY Adam Silverstein
ITALIAN LITERATURE
    Peter Hainsworth and David Robey
JESUS Richard Bauckham
JOURNALISM Ian Hargreaves
JUDAISM Norman Solomon
JUNG Anthony Stevens
KABBALAH Joseph Dan
KAFKA Ritchie Robertson
KANT Roger Scruton
KEYNES Robert Skidelsky
KIERKEGAARD Patrick Gardiner
THE KORAN Michael Cook
LANDSCAPES AND
    GEOMORPHOLOGY
    Andrew Goudie and Heather Viles
LANGUAGES Stephen R. Anderson
LATE ANTIQUITY Gillian Clark
LAW Raymond Wacks
THE LAWS OF THERMODYNAMICS
    Peter Atkins
LEADERSHIP Keith Grint
LINCOLN Allen C. Guelzo
LINGUISTICS Peter Matthews
LITERARY THEORY Jonathan Culler
LOCKE John Dunn
LOGIC Graham Priest
MACHIAVELLI Quentin Skinner
MADNESS Andrew Scull
MAGIC Owen Davies
MAGNA CARTA Nicholas Vincent
MAGNETISM Stephen Blundell
MALTHUS Donald Winch
MAO Delia Davin
MARINE BIOLOGY Philip V. Mladenov
THE MARQUIS DE SADE John Phillips
MARTIN LUTHER Scott H. Hendrix
MARTYRDOM Jolyon Mitchell
MARX Peter Singer
MATHEMATICS Timothy Gowers
THE MEANING OF LIFE Terry Eagleton
MEDICAL ETHICS Tony Hope
MEDICAL LAW Charles Foster
MEDIEVAL BRITAIN
    John Gillingham and Ralph A. Griffiths

MEMORY Jonathan K. Foster
METAPHYSICS Stephen Mumford
MICHAEL FARADAY
   Frank A. J. L. James
MODERN ART David Cottington
MODERN CHINA Rana Mitter
MODERN FRANCE
   Vanessa R. Schwartz
MODERN IRELAND Senia Pašeta
MODERN JAPAN
   Christopher Goto-Jones
MODERN LATIN AMERICAN
   LITERATURE
   Roberto González Echevarría
MODERN WAR Richard English
MODERNISM Christopher Butler
MOLECULES Philip Ball
THE MONGOLS Morris Rossabi
MORMONISM Richard Lyman Bushman
MUHAMMAD Jonathan A.C. Brown
MULTICULTURALISM Ali Rattansi
MUSIC Nicholas Cook
MYTH Robert A. Segal
THE NAPOLEONIC WARS
   Mike Rapport
NATIONALISM Steven Grosby
NELSON MANDELA Elleke Boehmer
NEOLIBERALISM Manfred Steger
   and Ravi Roy
NETWORKS Guido Caldarelli and
   Michele Catanzaro
THE NEW TESTAMENT
   Luke Timothy Johnson
THE NEW TESTAMENT AS
   LITERATURE Kyle Keefer
NEWTON Robert Iliffe
NIETZSCHE Michael Tanner
NINETEENTH-CENTURY BRITAIN
   Christopher Harvie and
   H. C. G. Matthew
THE NORMAN CONQUEST
   George Garnett
NORTH AMERICAN INDIANS
   Theda Perdue and Michael D. Green
NORTHERN IRELAND
   Marc Mulholland
NOTHING Frank Close
NUCLEAR POWER Maxwell Irvine
NUCLEAR WEAPONS Joseph M. Siracusa
NUMBERS Peter M. Higgins

OBJECTIVITY Stephen Gaukroger
THE OLD TESTAMENT
   Michael D. Coogan
THE ORCHESTRA D. Kern Holoman
ORGANIZATIONS Mary Jo Hatch
PAGANISM Owen Davies
THE PALESTINIAN-ISRAELI
   CONFLICT Martin Bunton
PARTICLE PHYSICS Frank Close
PAUL E. P. Sanders
PENTECOSTALISM William K. Kay
THE PERIODIC TABLE Eric R. Scerri
PHILOSOPHY Edward Craig
PHILOSOPHY OF LAW Raymond Wacks
PHILOSOPHY OF SCIENCE
   Samir Okasha
PHOTOGRAPHY Steve Edwards
PLAGUE Paul Slack
PLANETS David A. Rothery
PLANTS Timothy Walker
PLATO Julia Annas
POLITICAL PHILOSOPHY David Miller
POLITICS Kenneth Minogue
POSTCOLONIALISM Robert Young
POSTMODERNISM Christopher Butler
POSTSTRUCTURALISM Catherine Belsey
PREHISTORY Chris Gosden
PRESOCRATIC PHILOSOPHY
   Catherine Osborne
PRIVACY Raymond Wacks
PROBABILITY John Haigh
PROGRESSIVISM Walter Nugent
PROTESTANTISM Mark A. Noll
PSYCHIATRY Tom Burns
PSYCHOLOGY Gillian Butler and
   Freda McManus
PURITANISM Francis J. Bremer
THE QUAKERS Pink Dandelion
QUANTUM THEORY John Polkinghorne
RACISM Ali Rattansi
RADIOACTIVITY Claudio Tuniz
RASTAFARI Ennis B. Edmonds
THE REAGAN REVOLUTION Gil Troy
REALITY Jan Westerhoff
THE REFORMATION Peter Marshall
RELATIVITY Russell Stannard
RELIGION IN AMERICA Timothy Beal
THE RENAISSANCE Jerry Brotton
RENAISSANCE ART Geraldine A. Johnson
RHETORIC Richard Toye

RISK Baruch Fischhoff and John Kadvany
RIVERS Nick Middleton
ROBOTICS Alan Winfield
ROMAN BRITAIN Peter Salway
THE ROMAN EMPIRE Christopher Kelly
THE ROMAN REPUBLIC
    David M. Gwynn
ROMANTICISM Michael Ferber
ROUSSEAU Robert Wokler
RUSSELL A. C. Grayling
RUSSIAN HISTORY Geoffrey Hosking
RUSSIAN LITERATURE Catriona Kelly
THE RUSSIAN REVOLUTION
    S. A. Smith
SCHIZOPHRENIA Chris Frith and
    Eve Johnstone
SCHOPENHAUER
    Christopher Janaway
SCIENCE AND RELIGION Thomas Dixon
SCIENCE FICTION David Seed
THE SCIENTIFIC REVOLUTION
    Lawrence M. Principe
SCOTLAND Rab Houston
SEXUALITY Véronique Mottier
SHAKESPEARE Germaine Greer
SIKHISM Eleanor Nesbitt
THE SILK ROAD James A. Millward
SLEEP Steven W. Lockley and
    Russell G. Foster
SOCIAL AND CULTURAL
    ANTHROPOLOGY
    John Monaghan and Peter Just
SOCIALISM Michael Newman
SOCIOLINGUISTICS John Edwards
SOCIOLOGY Steve Bruce
SOCRATES C. C. W. Taylor
THE SOVIET UNION Stephen Lovell
THE SPANISH CIVIL WAR Helen Graham

SPANISH LITERATURE Jo Labanyi
SPINOZA Roger Scruton
SPIRITUALITY Philip Sheldrake
STARS Andrew King
STATISTICS David J. Hand
STEM CELLS Jonathan Slack
STUART BRITAIN John Morrill
SUPERCONDUCTIVITY
    Stephen Blundell
SYMMETRY Ian Stewart
TERRORISM Charles Townshend
THEOLOGY David F. Ford
THOMAS AQUINAS Fergus Kerr
THOUGHT Tim Bayne
TOCQUEVILLE Harvey C. Mansfield
TRAGEDY Adrian Poole
THE TROJAN WAR Eric H. Cline
TRUST Katherine Hawley
THE TUDORS John Guy
TWENTIETH-CENTURY BRITAIN
    Kenneth O. Morgan
THE UNITED NATIONS
    Jussi M. Hanhimäki
THE U.S. CONGRESS
    Donald A. Ritchie
THE U.S. SUPREME COURT
    Linda Greenhouse
UTOPIANISM Lyman Tower Sargent
THE VIKINGS Julian Richards
VIRUSES Dorothy H. Crawford
WITCHCRAFT Malcolm Gaskill
WITTGENSTEIN A. C. Grayling
WORK Stephen Fineman
WORLD MUSIC Philip Bohlman
THE WORLD TRADE ORGANIZATION
    Amrita Narlikar
WRITING AND SCRIPT
    Andrew Robinson

Available soon:

MANAGEMENT John Hendry
FRACTALS Kenneth Falconer
INTERNATIONAL SECURITY
    Christopher S. Browning

ENTREPRENEURSHIP
    Paul Westhead and Mike Wright
ASTROBIOLOGY
    David C. Catling

For more information visit our website
www.oup.com/vsi/

John Krebs

FOOD

A Very Short Introduction

OXFORD
UNIVERSITY PRESS

OXFORD
UNIVERSITY PRESS

Great Clarendon Street, Oxford, ox2 6DP,
United Kingdom

Oxford University Press is a department of the University of Oxford.
It furthers the University's objective of excellence in research, scholarship,
and education by publishing worldwide. Oxford is a registered trade mark of
Oxford University Press in the UK and in certain other countries

Published in the United States of America by Oxford University Press
198 Madison Avenue, New York, NY 10016, United States of America

British Library Cataloguing in Publication Data

Data available

ISBN 978-0-19-966108-4

Printed in Great Britain by
Ashford Colour Press Ltd, Gosport, Hampshire

Contents

Acknowledgements xi

Preface xiii

1 The gourmet ape 1

2 I like it! 21

3 When food goes wrong 43

4 You are what you eat 65

5 Feeding the nine billion 87

References 113

Further reading 123

Index 125

Acknowledgements

Many colleagues have helped me by providing information and steering me towards relevant literature. Susan Jebb and Sarah Phibbs both read a draft of the whole book and made valuable comments. Charles Godfray commented on chapter 5 and Andrew Wadge on chapter 4. The first draft was completed during a sabbatical term granted by the Fellows of Jesus College, Oxford. Latha Menon encouraged me to write the book, waited patiently for its appearance, and made very helpful editorial suggestions. I am grateful to all.

Preface

Everyone has a view about food, which is not surprising given that we eat in the region of 1,000 meals a year. Some people treat food simply as fuel, and do not take too much notice of what they are eating, but given that you have opened this book, you may be someone, like me, for whom food is a source of pleasure and fascination.

Think of the best dish you have ever eaten. Perhaps it was a chicken curry, or a delicious chocolate cake, or pasta with tomato sauce?

These dishes raise many questions about food. What is it about certain foods that makes them so delicious? Why is Indian food so spicy? Why do some people like spicy food while others don't? Why do so many people like chocolate? How is it that Italian food often contains tomatoes, a plant from South America? Why is rice the staple food for half the people on the planet? All the three dishes that I listed are prepared by cooking: when did cooking start?

I hope that you will find answers to these and many other questions as you read this book. As you read, you might wish to bear in mind that the question 'why?' can be answered in different ways.

For example, take the question 'Why do all humans tend to like fat, sugar, and salt?'

One way to answer is to talk about survival value. Fat and sugar, both crucial sources of energy, were important for survival in our evolutionary past, and there was therefore natural selection for an inbuilt preference for these components of our diet. Likewise, sodium and chloride ions, the components of salt, are essential constituents of body fluids and are lost each day in sweat and urine, so need to be replenished.

Another kind of answer refers to the sensory mechanisms that underlie our preference for these foods. We have taste receptors for sugar and salt on our tongue, and the 'mouth feel' of fat is rewarding. Our preferences are influenced by sensations in our mouth, but also by other senses such as sight, sound, and smell.

Still a third way to answer the question is in relation to how our childhood and social experience influence our food preferences. Sugary, fatty foods are often associated with treats for children. Perhaps part of our adult preference stems from these childhood experiences.

The book combines science, history, and culture. If it gives you a taste for more, it will have achieved its aim.

Chapter 1
The gourmet ape

Introduction

The traditional Inuit of the Canadian Arctic were almost entirely carnivorous. They hunted for seals, whales, birds as well as land mammals such as caribou, and ate only small amounts of plant food. Some traditional Inuit populations, it is estimated, obtained about 99 per cent of their energy intake from animal foods. On the other hand, approximately one-fifth of the people in the world never eat meat. For many this is because they cannot afford it, for some because their religious beliefs prohibit it, and for others because of perceived health, ethical, or environmental benefits. Most of us live between these extremes, eating a mixed diet that includes meat, dairy products, and plant-based foods.

The great variation between populations and individuals in the make-up of their diet raises the question of whether or not there is such a thing as a 'natural diet' to which the human body is adapted, or whether we are just extremely flexible. There are those who suggest that the best diet is the so-called 'paleo diet'. This is based on the argument that we are genetically adapted to the diet of our hunter-gatherer ancestors, because for about 99.5 per cent of human history we have existed as omnivorous hunter-gatherers, eating a mixture of fruit, nuts and seeds, roots, tubers, and a large variety of animals. But given that present-day humans appear to be able to survive well on such a wide range of foods, it seems more likely that our omnivorous past has left us with flexibility to cope with many different food sources. In this chapter I will explore the diets of ancestral hominins, our closest relatives in the fossil record, and highlight the major changes that have led to today's diets. To prepare the ground, let us begin with a very brief summary of the evolution of the human species.

Human evolution

Homo sapiens, modern man, first appeared in the fossil record between 200,000 and 250,000 years ago. The evolutionary history of humans both before and after the appearance of *Homo sapiens* has been pieced together from a small number of fossils of differing ages from different parts of the world. The fossils may be tiny fragments, such as the piece of bone, about 40,000 years old, from a fifth finger, discovered, along with a molar tooth, in the Denisova Cave in Siberia in 2008. Working out how the various fossil remains relate to one another is rather like trying to guess what the picture in a completed jigsaw would look like without having seen the cover of the box and with only a few random pieces laid out on the table.

There is a further complication. Not only is it a challenge to decide how different species of fossil hominins are related to one another, but also it is often difficult to determine whether or not different fossils belong to the same or different species. Some paleoanthropologists who study human origins are 'lumpers', tending to assign many skeletons and fragments to just a few species, whilst others are 'splitters', dividing the fossils into many different species. So different accounts of human evolution interpret the fossil record in very diverse ways.

Within the past twenty years or so, a powerful new tool has been deployed in the study of human origins: the analysis of ancient DNA recovered from fossil remains. The oldest useable DNA extracts are from human fossils are about 80,000 years old, but technical developments are pushing the time limit further back, especially for very well preserved remains. The degree of similarity between the DNA extracted from a fossil and that of modern man can be used to estimate how far back in time the fossil and modern man shared a common ancestor. For instance, the Denisova fossil, referred to above, shared a common ancestor with *H. sapiens* about 1,000,000 years ago, suggesting that it is a separate species. Similarly Neanderthal man, *Homo neanderthalensis*, is now known to have been a distinct species that separated from *H. sapiens* about 600,000–700,000 years ago.

To greatly simplify the story of our ancestry, most experts think that the direct precursor of *H. sapiens* was a species called *Homo erectus*, which first appeared about 1.8 million years ago. The taxonomic 'splitters' link *H. erectus* and *H. sapiens* with an intermediate species called *Homo heidelbergensis* that appeared about 800,000 years ago. *H. erectus*, although it had a smaller brain and would have been rather heavy-browed, looked sufficiently similar to *H. sapiens* for some anthropologists to have quipped, with slight exaggeration, that if you dressed *H. erectus* in modern clothes no one would notice him in the supermarket.

Further back in time, our ancestors included *Homo habilis* (nicknamed 'Handy Man' by its discoverer Louis Leakey, because it made stone tools), that lived in East Africa from about 2.4 million to 1.4 million years ago, and possibly *Australopithecus afarensis*, made famous by the skeleton of 'Lucy', that lived, also in East Africa, between three and four million years ago. The australopithecines, although bipedal, were very ape-like in appearance and were only about 1–1.3 m tall. If they were around today we might well see them in the zoo, alongside other, more distant relatives such as the chimpanzee. Many species of hominins existed, some in different parts of Africa, during this period and no one is sure which was the direct ancestor of modern man.

The diet of our ancestors

What did the australopithecines, and early species of *Homo*, eat? We can infer their diets from a combination of pieces of fossil evidence: teeth, tools, and chemicals. Although the digestive tract is not preserved in fossils, there are a few fossil skeletons of hominins that show indirectly how large the intestines were by the shape of the ribcage. Present-day great apes that eat low-quality plant food have a large digestive tract for absorbing nutrients and a large ribcage to accommodate their guts. I will return to this later when I discuss cooking.

The teeth of different species of mammals are clues to their diet. For instance, the cheek teeth, or premolars and molars, are quite different in a sheep and a dog: sheep have flat-topped molars with rows of ridges for grinding up tough grasses, whilst dogs have pointed molars for chewing meat. The front teeth, incisors and canines, tell a similar story. Sheep have sharp, flat-topped incisor teeth for cutting grass and have no canine teeth at all, while dogs have pointed canines and incisors that are well suited to tearing meat. The teeth of early hominins going back to 3.5 million years ago were neither the teeth of a pure herbivore, nor those of a pure carnivore, so they could have been the all-purpose teeth of an omnivore. It is not possible to go beyond this broad generalization and say what proportion of the diet was animal or plant, or to say which particular kinds of plants and animals were eaten.

The microscopic patterns of wear on the grinding surface of fossil teeth also tell a story about diet. Broadly speaking, wear patterns that are a complex mixture of tiny pits and scratches of different orientations, referred to as 'surface complexity', indicate a diet that included hard seeds, nuts, and other brittle items, whilst a wear pattern of many fine parallel lines, known as 'anisotropic wear', suggests a diet of tough vegetation such as grasses and sedges, ground by the teeth moving back and forth like a sheep's teeth.

Because the surface microwear patterns reflect only the most recent diet of the fossil individual, they do not reveal whether or not there were seasonal changes, or changes during the lifetime. Microwear studies of *A. afarensis*, for instance, suggest that it fed more on grasses and other tough vegetation than on seeds and nuts, but do not indicate whether or not it also ate meat.

Tools also tell us about diet and are one indicator that ancestral species of *Homo* ate meat as well as vegetable matter. The very earliest known stone tools were used by *H. habilis* about 2.5 million years ago for cutting flesh off bones. Fossil bones of non-human animals, such as zebras, discovered alongside *H. habilis* fossils, have grooves on them that were probably made by flint tools, suggesting that the stone tools were used to cut off meat. These early hominins could have been both hunters and scavengers, benefiting from large animals that had been killed by other predators such as lions. Whether hunter or scavenger, or both, *H. habilis* differed from its australopithecine forebears in that its diet probably included more meat. It is possible that tool use goes back even further: two bones of large ungulates 3.3 million years old, discovered in Ethiopia in 2009, have cut marks on them that appear to have been made by stone tools. If this interpretation is correct, *A. afarensis* was the first known tool user.

The kind of food an animal eats leaves a chemical signature in its tooth enamel. Plants and animals are largely made out of four chemical elements: carbon, hydrogen, oxygen, and nitrogen. Plants take up carbon, in the form of carbon dioxide, from the atmosphere and turn it into glucose in the process of photosynthesis (see also Chapter 5). But the carbon atoms in the atmosphere are not all identical: the different forms, or isotopes, differ very slightly in mass. Two particular isotopes, called ^{12}C and ^{13}C, are differentially absorbed by different kinds of plants as a result of subtle differences in the sequence of chemical reactions involved in photosynthesis. So-called C_4 plants, which convert carbon dioxide into a four-carbon compound during photosynthesis and include many grasses and sedges, tend to have higher ratios of $^{13}C/^{12}C$ (known as $\partial^{13}C$) than do C_3 plants, including most trees, shrubs, and herbaceous plants. The value of $\partial^{13}C$ can be measured in fossil remains and indicates what kinds of food our ancestors ate. Tooth enamel is more stable than bone during fossilization and is therefore used to make the measurements. In modern animals the $\partial^{13}C$ signature can be traced through the food chain. Grazing animals such as antelope have higher values than do browsers such as giraffe, and the predators that feed on grazers also have high values, reflecting the content of their prey. Hence a high value of $\partial^{13}C$ could mean either that a fossil hominin ate grasses and sedges, or that it ate animals that fed on these plants.

The carbon isotope ratios in the tooth enamel of three fossil hominins have been analysed: *Australopethicus africanus*, a slightly more heavily built and more recent relative of *A. afarensis* from South Africa, and two species of heavy-jawed hominins with large ridges on their skull indicating huge chewing muscles, *Paranthropus robustus* from South Africa and *P. boisei* from East Africa. The latter is known as 'Nutcracker Man' because it was assumed that its massive jaws and large teeth were used for cracking hard nuts and seeds. *A. africanus* lived between two and three million years ago and the two species of *Paranthropus* between 2.5 and 1.2 million years ago.

The results are rather surprising. Nutcracker Man was not a nutcracker at all. Its diet was 75–80 per cent grasses or sedges. Both the two South African species, in spite of the difference in the size of their jaws, obtained about 25–35 per cent of their diet from C_4 sources, either grasses or sedges, or animals that fed on these plants. The patterns of microwear on their teeth suggest that grasses were not a major source of food. So the most probable interpretation is that these hominins were omnivorous opportunists, rather than exclusively plant eaters. Consistent with opportunism, the $\partial^{13}C$ signature also varies greatly between individuals and within an individual on a vertical transect through a tooth, which records changes in diet through an individual's life.

The detective investigation to work out the diets of fossil hominins still has a long way to go, but so far we can tentatively conclude that our evolutionary heritage, from the australopithecines to *Homo*, is of a flexible, opportunist omnivore with a gradual transition from plant- to animal-based diet. Hunting for meat is thought to have increased in importance during the later stages of human evolution.

During the journey from our hunter-gatherer ancestors to the present day there have been three seismic changes in the food we eat. The first was the discovery of cooking, the second was the emergence of agriculture, and the third, the invention of methods of preserving and processing food. A typical day's food for most people in the world today will include food that is cooked, farmed, and processed.

Cooking

Charles Darwin thought that cooking, after language, was the second greatest discovery made by man. There are people who believe that eating raw food is healthier than eating cooked food and there is even a 'raw food movement'. All of us eat some raw food: for instance fruit, vegetables, and sushi. But the great majority of foods we eat are cooked, and the fact is that, by and large, cooked food is easier to digest, safer, and more nutritious. It is very difficult for modern humans to remain healthy on a diet of purely raw food, even though the vegetables, grains, and fruit available today are far more digestible and nutritious than were the wild ancestors of these crops.

The discovery of cooking transformed the food lives of our ancestors. The anthropologist Richard Wrangham has argued that it had profound impacts not just on our diet, but also on our anatomy, brain, and social life.

Cooking can make plants that are inedible into edible food by destroying the poisons that plants often manufacture to protect themselves against attack by insects or other herbivorous animals. Kidney beans, for example, are toxic unless they are cooked, because cooking breaks down the toxin, lectin phytohaemagglutinin. These toxic chemicals are referred to as 'plant secondary compounds', because they are not directly involved in the plant's normal growth, development, and reproduction, and are produced purely as chemical defences. They give many of the plants we consume, such as coffee or Brussels sprouts, their bitter taste. I will return to the taste of bitter foods in Chapter 2 and the possible nutritional benefits of plant secondary compounds in Chapter 4.

Cooking also makes food safer by destroying harmful microorganisms such as *Salmonella* that can lead to a serious bout of food poisoning. Even those who advocate a raw food diet probably drink pasteurized milk that has been 'cooked' by heating it to about 70° C for fifteen seconds to kill bacteria. As we shall see in Chapter 3, pasteurization has saved many lives.

Cooked food is often more digestible and yields more energy, because heat breaks down tough cellulose cell walls in plants or tough connective tissue in animals. Chewing a raw turnip, a plate of uncooked rice, or a raw leg of lamb is much harder work than eating the cooked equivalent. The energy expended in chewing to break down the tough material is replaced by energy from the fossil or other fuel that is used in cooking the food, so the ratio of energy gained to energy expended by the body is greater when food is cooked.

It is sometimes said that raw vegetables are more nutritious, because some nutrients, such as vitamin C, are destroyed by cooking. But the fact that cooking also breaks down the indigestible cellulose cell walls of plants may make the food more nutritious by releasing nutrients from inside the cell that would otherwise be inaccessible to us.

The anthropologist Richard Wrangham has suggested that cooking led to the evolution, over many generations, of changes in our ancestors' anatomy, for instance a reduction in the size of teeth and jaws and shortening of our digestive tract, indicated in fossil remains by the decrease in size of the ribcage. Importantly, cooking also increased our energy intake by making food more easily digestible, and Wrangham thinks that this could have been crucial in the steep increase in brain size of our ancestors that started about 1.5–2 million years ago, because the brain is energetically very expensive to maintain. Our brain is 2 per cent of our body mass but accounts for 20 per cent of our resting energy expenditure. Australopithecines had a brain size of about 450 cc and *H. habilis* about 600 cc, whilst *H. erectus* had a brain measuring about 1100 cc, much closer to *H. sapiens* with a brain of 1330 cc. Even allowing for an increase in body size, the brains of *H. erectus* and *H. sapiens* are relatively substantially larger than those of their predecessors. Finally, Wrangham suggests, cooking enhanced our social life by freeing up time spent chewing for time spent chatting.

This hypothesis about the dramatic impact of cooking on human evolution is persuasive, but it is also controversial. The problem is that it has to assume that cooking was discovered about 1.9 million years ago, because the anatomical changes described above, including the increase in brain size, started then. However, the archaeological evidence of *systematic* use of controlled fire dates back to only about 250,000 years, although there is scattered evidence of fire associated with hominin settlements that are much older. One example is the remarkable discovery of more than half a dozen 400,000-year-old wooden throwing spears at Schonningen in northern Germany. The spears were found along with the remains of many horses, probably killed by the hunters for food, and there are signs of fire at the site. In Israel, archaeologists have discovered an 800,000-year-old hominin site with signs of fire and a hearth, around which various activities were carried out, such as making flint tools, roasting seeds, and eating fish, crabs, and a variety of mammals. There may be as yet undiscovered remains of controlled fire from even earlier times, such as the 1.0–1.5 million-year-old charred bones discovered at Swartkrans in South Africa. The fossil record of human evolution is very sparse, and during the past few decades many important new finds have dramatically altered our view of human evolution. Perhaps our distant ancestors, at least the very few fossils of them so far discovered, did not die close to the places where they made fires.

The fact that during the past 1.8 million years the anatomy and physiology of our ancestors changed dramatically in ways that make us suited to eating cooked food is not in doubt. Our jaws are too weak, our teeth and mouth too small, and our digestive tracts too short to cope with wild raw food, and our brains need the rate of energy intake that cooked food provides. Many of the foods that our closest primate relatives eat are too tough, too toxic, and too bitter for us to swallow, and the raw meat carcasses that non-humans can eat would give us food poisoning.

This loss of the ability to live healthily on raw, wild food is an example of the way in which evolution by natural selection pares away unnecessary parts of an organism's anatomy and physiology. The dodo lost the ability to fly on an island with no predators and the cave tetra fish has lost its pigmentation and eyes. This happens for two reasons. First there is no such thing as a free lunch: manufacturing and maintaining body parts cost energy and carry risks such as infection or malfunction, so if the parts do not enhance survival, selection will eliminate them because of their costs. Second, and linked to the first, there may be a positive advantage in allocating resources during growth and development to other functions, for instance the pressure sensors that enable the blind tetra fish to 'see in the dark'.

So the changes in the anatomy and physiology of *Homo erectus* must have happened because the large teeth, jaws, and guts were no longer advantageous. If the evolutionary driver of this change was not the discovery of cooking, it must have been something else. The major alternative explanation is that the changes were driven by the switch to a diet based on high-quality animal food and the use of tools to cut meat into small pieces to make it more easily digestible. A possible argument against this idea is that, while it is true that carnivores tend to have larger brains relative to body size than do herbivores, no other carnivorous species has changed so dramatically in evolution as have hominins. However, it is possible that a shift towards meat and cooking both contributed to the changes in hominin anatomy and physiology: the two ideas are not mutually exclusive.

Agriculture

The few remaining hunter-gatherer groups in the world today, in Africa and South America, spend up to seven hours a day gathering food. It is a time- and energy-intensive way of making a living. Hunter-gatherers also live in small groups and at low population density because they need a large area to gather enough food for survival. When our ancestors were hunter-gatherers, the total world population of humans was probably no greater than that of London or New York today.

This all began to change around 10,500 years ago, when, in several places in the world, humans started farming. Although there is some debate about whether it had five, seven, or more independent origins, agriculture appeared over a period of a few thousand years, probably independently, in Southwest Asia (present-day Iraq), China, Africa, New Guinea, North, Central, and South America. Different crops and livestock were domesticated in different places. In Southwest Asia, for instance, people farmed olives, wheat, peas, goats, and sheep, while in Central America they farmed maize, beans, squash, and turkeys. It seems plausible that in each area, hunter-gatherers chose the crops and animals from the foods they were already eating that

11

were most suited to domestication. Perhaps crops were originally planted by accident, when seeds and fruit that had been gathered in the wild were dropped and grew around small human settlements. Seeds often pass through the gut and come out at the other end able to germinate, and this is their natural dispersal mechanism. So latrines near human settlements could have been the fertile ground for the start of plant domestication. Some animals that were hunted may have been relatively easy to keep near home, and eventually were bred systematically. Once agriculture started, it spread around the world rapidly, presumably because agricultural populations reproduced more rapidly than hunter-gatherers.

The earliest farmers were also the first, albeit unwitting, geneticists. By selecting, over many tens, hundreds, or thousands of generations, the seeds with the biggest kernels or fewer of the plant toxins that make them unpalatable, the sheep that produced the most milk, or the pigs that grew quickly, the prehistoric farmers gradually turned the wild ancestral species into domesticated agricultural crops and livestock. The evolutionary change of domesticated species brought about by farmers was used by Charles Darwin to support his theory of natural selection. He coined the term 'artificial selection' when he wrote, in the *Origin of Species*:

> Slow though the process of selection may be, if feeble man can do much by his powers of artificial selection, I can see no limit to the amount of change, to the beauty and infinite complexity of the coadaptations between all organic beings, one with another and with their physical conditions of life, which may be effected in the long course of time by nature's power of selection.

Much of the early development of the science of genetics in the first half of the 20th century was driven by the wish to better understand how artificial selection can be used to improve agriculture. I will return to this theme in Chapter 5.

Although we will never know how domestication started, we can say a number of things with a degree of confidence.

Out of the many thousands of plant species that could in theory have been domesticated, only very few have come to dominate world agriculture. Among cereal crops, for instance, just three—maize, rice, and wheat—account for 87 per cent of all grain production worldwide and 43 per cent of all calories consumed, according to the United Nations Food and Agriculture Organization. Six more crops—barley, sorghum, oats, millet, rye, and triticale (a hybrid of wheat and rye)—account for most of the rest. Leguminous crops, including beans, peas, and lentils, are similarly dominated by a very small number of species, as are farmed animals. Jared Diamond, in his book *Guns, Germs and Steel*, estimates that out of 148 large herbivorous mammals that could in theory have been domesticated, only fourteen were chosen. As with plants, only a small number of species—sheep, goats, cattle, and pigs—dominate world food production. The great majority of farmed birds are chickens, turkeys, ducks, and geese.

Diamond's answer to the question of why so few species have been domesticated out of so many candidates is, in a word, suitability. For example, the animal species that our ancestors homed in on might have been chosen for a combination of factors such as size, tameness, growth rate, diet, and social organization (highly territorial species would not be suitable!). One could construct an equivalent set of criteria for plant crops, including digestibility, ease of cultivation, and growth rate, to explain why so few of the thousands of potentially suitable plants have been domesticated.

According to Diamond, agriculture paved the way for dramatic changes in human societies, including new technologies, increased population size, larger communities, and the possibility of specialization in different tasks, such as farming and fighting, within society. These changes in turn led to the spread of diseases, the development of writing as a way of keeping records, formalized government structures, and eventually to modern nation states.

At least in the short term, the net effect of agriculture on human health seems to have been negative. The earliest farmers were shorter than their hunter-gatherer ancestors, and they suffered from tooth decay, probably because they were eating carbohydrates from the crops they were growing. These changes were almost certainly not the result of genetic alterations in the human population, but simply an effect of nutrition resulting from a less varied, and perhaps lower protein, diet. But the adverse effects of farming on health may have moderated over time. Perhaps the most detailed evidence for this comes from a study by Anne Starling and Jay Stock of 242 individuals from the Nile Valley between 13000 and 1500 BC. The pre-agricultural skeletons were 10 cm taller (1.73 m) and had fewer signs of poor nutrition, indicated by loss of tooth enamel. Among early farmers, from 4000 to 5000 BC, 70 per cent were not well nourished compared with only 39 per cent of their hunter-gatherer ancestors. But by 2000 BC the agriculturalists were better nourished than the hunter-gatherers, with only 20 per cent showing signs of enamel loss. So it appears that as agriculture became better established and more efficient, it provided a more reliable food supply than did hunting and gathering.

Farming has survived and spread primarily because we have, for the past 10,000 years, enjoyed a remarkably benign and stable climate. For all of human history (including the precursors of our own species such as *H. habilis* and Australopithecines) the climate of the Earth has oscillated between cold (glacial) and less cold (interglacial) periods. The details of these changes are recorded as chemical and physical changes in ice in Antarctica. By drilling very deep cores into the ice, the deepest of which, the Vostok core, is more than 3 km long, the 'fossil record' of the Earth's climate has been analysed over the past 400,000 years in great detail. The

14

last glacial period ended about 11,000 years ago. The whole of agriculture, and associated development in human civilization, has been crammed into the current interglacial period. These warm periods typically last for between 10,000 and 30,000 years, so in theory we are due, at some stage in the distant future, for a very long, very cold, period, unless man-made global warming is sufficient to override the natural cycle. In Chapter 5, I will return to the question of the future of food production, including the effects of climate change, whether natural or man induced.

Preservation and processing

If you consider the foods you have eaten in the past day or two, the chances are that many of them were preserved in one way or another, or processed to transform the raw ingredients into an edible, even delicious, composite of different components. Fruit juice or milk will probably have been pasteurized; the wheat or rye grains of the bread ground into flour and mixed with yeast, water, and salt before baking; jam preserved by adding sugar and cooking; and the tea leaves will have been dried.

The arrival of agriculture meant that, for the first time, our ancestors had more food than they could eat immediately. This, combined with the seasonality of production, led them to discover methods of preserving food: smoking (usually combined with salting and drying), drying, adding acid by fermentation, or adding salt. We do not know exactly when and where these methods developed, but they were probably in use soon after the dawn of agriculture, and discovered independently in more than one place, because different methods were appropriate according to the kind of food and the locality. For instance, drying works well for fruits in a hot sunny climate and salting with sea salt is suitable for preserving fish in coastal areas.

The four methods all share one feature in common: they make the food a more hostile environment for bacteria, moulds, and other microorganisms that cause food to go rotten. They also tend to slow down any natural chemical reactions in the food that would cause decay. Although our ancestors did not know about microorganisms, they must have discovered by trial and error what helped to prevent decay of food. Leave a bunch of grapes out for two weeks and they will be a mouldy mess, but dried raisins will be fine. Legs of air-dried Serrano ham hang happily in Spanish tapas bars for weeks on end, while a leg of raw pork would go off in a few days. Long after the milk in the fridge has gone sour, the piece of hard cheese bought on the same day will be fine.

Pickled food in some cases is made by adding vinegar, itself a product of fermentation, and in others by creating an environment, by adding salt, in which bacteria produce lactic acid to act as a preservative. This is how sauerkraut and its Korean cousin kimchi are produced. One of the most important bacteria used in manufacturing fermented dairy products such as yoghurt or cheese, *Streptococcus thermophilae*, has been shown, by analysing its DNA sequence, to have evolved from a related pathogenic (harmful) bacterium that infects the human mouth, *S. salivarius*. The genetic information suggests that the fermentation bacterium may have arisen about 10,000 years ago, when our ancestors first started to domesticate animals. During its evolution, *S. thermophilae* has lost the genes that were responsible for its ability to cause harmful infection to humans. The same seems to have happened with other fermentation bacteria belonging to the genus *Lactobacillus*. The intriguing hypothesis that these observations suggest is that fermentation of milk may have started when people accidentally spat into drinking vessels. However, not all microorganisms that cause fermentation have come from within the human body. The yeasts that create the alcohol in wine and beer occur naturally on raw ingredients, such as grapes and grains.

According to Pliny the Elder, Roman soldiers were paid in salt, giving rise to the word 'salary'. Phrases such as 'below the salt', 'worth his salt', or 'the salt of the earth' give us a hint that salt was an important commodity in medieval times. In northern Europe, it was expensive, and only afforded by the wealthy, because it was imported from southern Europe where it was made by evaporation of seawater, a process that still survives in many parts of the Mediterranean. Salting is one of the very ancient methods of preserving food, but it became particularly important in Europe in the late Middle Ages for preserving fish.

Although today's shopper still buys many foods preserved in the ancient ways, two much more recent methods of preserving food have come to predominate on supermarket shelves: canning and freezing.

The invention of canned food is a remarkable example of how technology precedes scientific discovery.

Canning was invented by a Frenchman, Nicholas Appert, in the first few years of the 19th century. He sealed food in glass bottles and then heated them in boiling water to cook the contents, a process known as appertization. In effect, Appert was pasteurizing the food, although Louis Pasteur did not discover that heating kills the microorganisms that spoil food until 1862. Appert developed his technique in response to a competition. The aphorism 'An army marches on its stomach' ('C'est la soupe qui fait le soldat') is attributed to Napoleon Bonaparte, and in 1795, the French government offered a prize of 12,000 francs for the person who could feed Napoleon's army on the move by preventing food spoilage. Appert was awarded the prize in 1809 with the condition that he published his method, which he did in his 1810 book *The Art of Preserving all Kinds of Animal and Vegetable Substances for Several Years*.

Appert's method had great advantages over older methods of food preservation: it could be applied to a wide range of foods, and the flavour and texture of the food were not too different from the freshly cooked product. His idea was soon copied by an Englishman, Peter Durand, who replaced Appert's glass bottles with tin cans, which had the advantages over glass of weight and resistance to damage. Two years later, in 1812, two Englishmen, Bryan Donkin and John Hall, started the commercial canning of food, although the real take-off in popularity of canning had to wait until the can opener was invented in 1855. Up to this time, cans were opened with a hammer and chisel! Donkin and Hall's cans were handmade and a skilled workman could produce about six per hour, compared with a modern factory that makes 1,200 per minute.

Canning is an extremely effective way of preserving food: a can of meat and gravy dating from 1824 was opened in 1939 and the contents were still in good condition. Cans are now used for drinks, pet food, and many non-food products. But in recent decades their dominance in food preservation has, at least in rich countries, been to a considerable degree eroded by another technology: freezing.

Chilling food to keep it fresh is an old idea. The earliest records of icehouses (called Yakhchals), thick-walled buildings, half underground, date back to 1700 BC in northwest Iran. In early 16th-century Italy, water was mixed with salt to lower its freezing point to −18°C, and by the mid 19th century frozen fish and meat were transported by ship from Australia to England. But the modern frozen food industry was started in the 1920s, by an American, Clarence Birdseye. Birdseye's key discovery, which he made as a result of observing the Inuit on a fishing trip to the Canadian Arctic, was that very rapid freezing creates smaller ice crystals and therefore causes less damage to the food. That is why his frozen food was better than that of his competitors. Nevertheless, the big growth in demand had to wait for the arrival of deep freezers, or refrigerators with freezer sections, in the home. The advantages of frozen over canned food include the fact that the flavour and texture are often indistinguishable from the equivalent fresh product, and that freezing can be used to preserve a huge variety of foods, from cakes to cauliflower, chicken curry to chocolate mousse.

Food processing

Cooking and preservation are two important elements of food processing. Processing may also involve mixing ingredients, or transforming their physical or chemical properties. Bread is a very familiar and ancient example of a compound processed food, probably originating in the Middle East, where the most suitable, gluten-rich cereals such as wheat were first domesticated, at around the time of the start of agriculture. The earliest breads were unleavened, consisting of ground grain and water cooked on a flat stone, and the discovery of leavening by yeast, well established in ancient Egypt 4,000 years ago, was almost certainly accidental as a result of using cereals that were infected with natural yeasts.

Four elements are needed to make unleavened bread: water, a gluten-rich grain (the gluten causes the bread to hold together), a mortar and pestle for grinding the grain, and a hot surface. Add yeast and an oven, and you are not far from a modern loaf. In contrast, another early processed food, chocolate, requires many steps to turn something that is totally inedible into a highly desirable food or drink, and the method of processing has evolved gradually. It is a remarkable illustration of the progressive development of food processing over many centuries, and a striking example of the unique ability of *H. sapiens* to turn the inedible into the edible.

Chocolate is made from the seeds of the cacao tree *Theobroma cacao* (Theobroma means 'food of the gods'), a native of Central and South America. The seeds, or beans, grow inside an oval fleshy pod that looks a bit like a melon. Between 2,000 and 1,000 years ago, Mayans in Central America fermented cacao beans with their surrounding sugary pulp in the sun. The dried beans were then winnowed, roasted, and ground before mixing with hot water and flavourings such as chilli and vanilla, and wild flowers. Drinking cacao was an important part of ceremonial banquets; the beans were a valuable, traded commodity, and in some places served as a currency.

Cacao made its way to Europe in the 16th century. The name chocolate is thought to be derived from the Mayan word 'chocol' (hot) and the Nahuatl word 'atl' (water).

Today's chocolate is very different from the drink of 17th- and 18th-century chocolate houses of Europe, which would seem to us to be fatty and gritty. The kernel of the cacao bean is about 50 per cent fat, and it was not until 1828 that Casparus van Houten invented a screw press for squeezing out much of this fat. The resulting 'cake', called cocoa powder, was similar to modern cocoa powder: it mixed more readily with sugar and water or milk and it was less fatty than its predecessor.

Van Houten also realized that by mixing some of the cocoa butter back into the powder, he could make a solid chocolate bar, nowadays stirred for many hours or days during processing to create a smooth texture.

In the next chapter I will explore likes and dislikes, and chocolate is a good starting point. Why is chocolate so appealing? There have been many speculations. For instance, chocolate contains tiny quantities of cannabinoids and amphetamine-like chemicals, which some have argued make chocolate addictive. Chocolate also contains small amounts of tryptophan, which is converted in the body to serotonin, a neurotransmitter in the brain thought to be associated with 'happy feelings'. More straightforwardly, fat and sugar, the main ingredients of chocolate, are inherently appealing to us because of their high energy content. Chocolate also turns from a solid to a liquid at mouth temperature: it literally melts in the mouth.

Chapter 2
I like it!

Introduction

Oxford, England, is a city with a population of about 120,000. When it comes to eating out, there are Caribbean, Chinese, French, Greek, Lebanese, Indian, Italian, Japanese, Malaysian, Mexican, Moroccan, Spanish, Russian, Turkish, and Thai restaurants, not to mention numerous fast-food chains as well as traditional British pubs.

The mere fact that I have listed all these different kinds of restaurant is a reminder that, traditionally, cuisine, the food ingredients, and the ways in which they are prepared vary enormously from country to country and from culture to culture. We associate Indian food with fragrant spices such as cumin, coriander, fenugreek, cardamom, cinnamon, and turmeric, as well as the heat of chilli peppers. Japanese food, in contrast, brings to mind sushi, soy sauce, ginger, and wasabi, whilst Italian food conjures up pasta and the flavours of tomato and garlic. The range of restaurants on offer in Oxford is vastly greater than it was fifty years ago, and shows how readily we modify our likes and dislikes. Globalization, travel, and affluence have brought to the rich countries, and increasingly to the developing world, an incredible diversity of food, and people have lapped it up. There may, however, be limits: Oxford has not yet seen a rush to open restaurants serving the southeast Asian speciality balut, boiled duck embryo inside the egg, or a drink made from cows' urine consumed by certain Hindus.

How did variation in cuisine originate? Why do we like some foods and not others? Are people in different parts of the world born with different kinds of taste preferences? Do people simply depend on the food that can be found locally? Do we learn to like what we are given as children and avoid the unfamiliar? Are our preferences a result of cultural or religious traditions?

Of course these, as well as other hypotheses, are not mutually exclusive. For instance, a tradition could become a tradition precisely because it reflects the food that is locally available, and furthermore, the food we are fed when we are young has a big influence on our later preferences, which is probably the main way in which traditions for cuisine are passed from generation to generation.

Taste

Let's start with the basics: the sensory experience when we eat an item of food. Although one might at first think of food preferences as largely to do with taste, there is much more to it than that. What we refer to as the taste of a food is actually often its flavour, which is the result of more than one sensory input, as explained below. Five different kinds of taste receptor on the tongue have been identified to date: sweet, sour, bitter, salt, and 'umami', which detects savoury foods. These five tastes are part of our basic ancestral survival kit for eating: sweet for selecting energy-rich foods, bitter for avoiding poisons, sour for avoiding foods that are decomposing, salt because the body needs to replenish the salt lost each day in sweat and urine, and umami for detecting proteins that are essential building blocks of the body.

The idea that there are a small number of tastes goes back to the ancient Greeks, as well as to some Eastern traditions. However, in the past few decades scientists have analysed the basic biological mechanisms of taste: how individual receptor cells on the tongue react with the chemicals in food to generate the nerve impulses that we interpret as taste. The molecular structure of, and genetic variation in, these receptors is still the subject of active research, and it is possible that other kinds of taste receptor will be discovered. For instance there have been some reports of a taste receptor for fat. As an aside, the 'taste' of 'hot', as in chilli or black pepper, is also detected in the mouth, but by pain receptors rather than taste receptors. Anyone who has rubbed their eyes after handling a scotch bonnet or bird's eye chilli pepper will recognize that we do not 'taste' the heat in chilli, but experience pain.

Umami

The taste receptor for umami is the one that has been identified most recently. It was discovered at the beginning of the 21st century, almost a hundred years after the taste had first been proposed in 1908 by a Japanese chemist, Professor Kikunae Ikeda. He discovered that two kinds of molecules, glutamate and certain ribonucleotides, give meat and other foods their savoury taste, and he coined the word umami, translated as 'yummy', to describe the taste. He also obtained a patent for the production of monosodium glutamate (MSG) as a flavour enhancer. Initially he extracted MSG from kelp (*Saccharina Japonica*) (Japanese konbu), but nowadays the two million tons per year of production is based on bacterial fermentation. MSG is widely used as a flavour enhancer, particularly in processed or preserved food, where the processing or preservation reduces the natural flavour.

There have been many reports about the adverse health effects of MSG, including the so-called 'Chinese Restaurant Syndrome' which includes headaches, chest pain, numbness or burning around the mouth, and a sense of facial pressure or swelling.

Since glutamate occurs naturally in the body as a chemical messenger between nerve cells, or neurotransmitter, there could in theory be an argument that MSG has some effects on the nervous system, although the link between MSG and Chinese Restaurant Syndrome has not been demonstrated. Furthermore, naturally occurring glutamate is eaten in many different foods without appearing to induce any adverse effects. There are especially high levels in Parmesan cheese and soy sauce, accounting for their intense savoury taste, and moderately high levels in walnuts, mushrooms, tomatoes, and many other foods, not to mention human breast milk. In addition, many processed foods that claim not to contain MSG often contain other flavour enhancers such as hydrolysed vegetable protein, which in turn contain glutamate. The use of glutamate as a flavour enhancer is not new. For instance, the Romans, without knowing about glutamate or umami, used a fish sauce called garum, akin to nam pla, the oriental fish sauce, for flavouring much of their food.

From taste to flavour

The five tastes do not explain the huge variety of flavours of food. The difference between a raspberry and a strawberry, between lamb and chicken, between a cinnamon roll and an almond croissant, are very largely unrelated to the differences in the five kinds of taste. This is because flavour includes a combination of the taste detected on the tongue and aromas detected with the nose. Hence people often report that food loses its flavour when they are suffering from a bad cold and have a blocked nose.

Simple home 'experiments' can reveal the roles of taste and aroma in creating flavour. The flavour of chewing gum is a result of its mint aroma and its sweet taste, generated either by sugar or by a sugar mimic such as aspartame that triggers the same receptor. When chewing gum loses its flavour is it the sugar, the mint, or both, that has 'run out'? The answer is that it is the sugar: adding a small amount of sugar to the mouth causes the mint flavour to reappear. Alternatively, breathe in, and then pinch your nose while chewing on a banana. While you hold your nose the banana is flavourless, but when you release your nose and breathe out you experience a flood of banana flavour, showing that this is an aroma and not a taste detected on the tongue.

This experiment also reveals that the aroma of banana is detected when you breathe *out*. This is because the smell that contributes to flavour is so-called retronasal scent, that is to say it is detected not by inhaling scent-laden air through your nostrils, but because the scent moves to the rear of your nasal passages from the back of the mouth where the two passages connect. The banana aroma, made up of more than 300 volatile chemicals, is detected by receptors in the nasal cavity.

It is often said that the human sense of smell is not as rich or acute as that of many other mammals, because we are diurnal and visual, while most mammals are nocturnal and rely much less on vision. Our sense of smell is nevertheless pretty impressive. We have just five kinds of taste receptor, but around 400 different kinds of receptor for smell. While this is fewer than the 1,000 found in rats, the neurobiologist Gordon Shepherd has argued that the greater complexity of the human brain for synthesizing and combining information from different receptors may well compensate for the smaller number of distinct kinds of reception.

In detecting a particular odour, the 400 receptors work in combinations and permutations, so that the number of different odours we can, in theory, distinguish is enormous. This is a great boon to the perfume industry. Look at the duty-free shop at a major airport: there are likely to be more than 500 varieties, and this is only a small proportion of the total. Our ability to distinguish a vast number of different odours is sometimes likened to our ability to recognize a very large number of colours. Colour is a creation of the brain, based on inputs from just three kinds of receptors, for red, green, and blue wavelengths, very loosely analogous to the creation of the illusion of a very large number of colours in a pointillist painting. Aroma is similarly a creation of the brain.

Receptors, for taste or smell, are located in the membrane that surrounds a cell. Three of the five taste receptors—bitter, sweet, and umami—act rather like a lock and key. Each receptor type is able to detect a particular molecule, and when the appropriate molecule fits into the receptor, the result is a cascade of molecular changes in the cell membrane and inside the cell itself, the end result of which is an electrical signal transmitting information to the brain or other parts of the nervous system. The taste for salt and acid results from specific kinds of change in molecule-sized pores called ion channels in the cell membrane, again producing a cascade of chemical changes that eventually produce an electrical signal. The receptors involved in scent detection are able to detect a range of different, related, chemicals, just as a particular kind of receptor in the eye involved in colour vision can detect a range of wavelengths of light.

But there is much more to the flavour of food than taste and smell. The brain integrates input from different senses, including sight, sound, temperature, pain, and mechanical sensations in our mouth such as crunchy or chewy, to create the overall sensory experience of eating food. When stimulation of one sensory pathway elicits a response in another, the phenomenon is called synaesthesia. For instance, some people when hearing a sound or a number experience the sensation of a particular colour. The interactions between different senses that determine flavour may be a kind of synaesthesia. Seeing a red strawberry may elicit a sensation of sweet taste. There is a debate among sensory psychologists as to whether 'flavour' should be treated as a separate sensory modality or a kind of perceptual system that integrates many different kinds of sensory information to create the equivalent of a picture.

Psychologists have investigated the ways in which different senses affect our experience of food. In one study, participants had to bite into 180 potato crisps (potato chips in North America) and rate them for crispness and freshness. The participants wore headphones through which they heard crisp-eating noises, and they rated the crisps as crisper and fresher when the noise was louder or emphasized higher frequencies. This vindicates an old cartoon in which a dejected inventor is standing in front of the boss's desk with the caption 'I am sorry, Smith, but the world is not ready for the silent crisp.' Background music, too, affects the experience of flavour. When volunteers ate toffee while listening to low-pitched, sombre brass music, they rated the toffee as more bitter than (the same) toffee eaten with a background of higher-pitched piano music. The colour, weight, and shape of cutlery also affects the perceived flavour of food.

Although it is obvious that part of our assessment of food is its visual appearance, it is perhaps surprising how visual input can override taste and smell. People find it very difficult to correctly identify fruit-flavoured drinks if the colour is wrong, for instance an orange drink that is coloured green. Perhaps even more striking is the experience of wine tasters. One study of Bordeaux University students of wine and wine making (oenology) revealed that they chose tasting notes appropriate for red wines, such as 'prune, chocolate, and tobacco', when they were given white wine coloured with a red dye. Experienced New Zealand wine experts were similarly tricked into thinking that the white wine Chardonnay was in fact a red wine, when it had been coloured with a red dye.

These kinds of finding have been put into practice both by the food industry to enhance the attractiveness of their products and by chefs who call themselves 'molecular gastronomists'. Oysters, for instance, are reported to be more enjoyable when accompanied by seaside sounds such as waves crashing on the beach than when accompanied by farmyard chicken sounds, and this has led the English chef Heston Blumenthal to develop a seafood dish in

I like it!

which the diner listens to seaside sounds through headphones while eating.

Which parts of the brain are involved in integrating all the information that creates a flavour? The answer is that many different parts are involved, including regions that process information from the different sensory inputs such as taste, smell, texture, sound, colour, temperature, and pain, areas of the brain concerned with memory, emotion, language, and hunger or satiety.

Sensory-specific satiety

If laboratory rats are fed on a monotonous diet of rat pellets they do not gain weight, but if they are fed on pellets of different flavours, they begin to get fat, because when there is variety they eat more. This is known as sensory-specific satiety. It is as though the rat gets bored with eating a monotonous diet, as do humans. Many of us will have had the experience of helping ourselves at a buffet and finding that we can eat much more than we would if we were just eating a single kind of food; and children sometimes say, after they are too full to eat any more chicken and peas, that their 'pudding tummy' could cope with a few scoops of ice cream. Even our favourite foods become intolerable if we eat too much of them: for instance, chocoholics lose interest in chocolate if they eat several bars in a row. As we shall see in Chapter 4, one factor contributing to the dramatic increase in obesity worldwide in the past few decades is the increased variety of food available, which overcomes our inbuilt sensory-specific satiety mechanisms.

Genetic differences in taste

These generalizations about flavour, taste, and other sensory experiences do not mean that everyone has the same sensory experience. Genetic differences affect our sense of taste, smell, vision, and other senses.

Genetic variation in our ability to taste bitter substances is the best-studied variation in taste. This was discovered by accident in 1930 by Arthur L. Fox, who worked for the Dupont Chemical Company. He found that people varied in their response to a bitter chemical called phenyl thiocarbamate (PTC) when it was dabbed on the tongue. At a meeting of the American Association for the Advancement of Science in 1931, he tested out the audience. He found that 28 per cent could not taste PTC, 65 per cent could, and the rest were not classifiable. Subsequent studies have divided the tasters into 'supertasters' who react with disgust when PTC is dabbed on their tongue, and 'tasters', who react with mild aversion. Tasters make up about half the population and supertasters 25 per cent, leaving the remainder as non-tasters. It is now known that this variation involves at least twenty-five different genes.

The receptors on the tongue that detect bitter tastes respond mainly to the bitter, poisonous compounds that plants produce as deterrents against herbivores, already referred to in Chapter 1. Brassicas such as cabbage, broccoli, and Brussels sprouts get their bitter edge from a family of compounds called glucosinilates, whilst the mouth-puckering bitterness in red wine and tea is caused by phenolic compounds, especially tannins. Supertasters are especially sensitive to these plant defences, so it is not surprising that they tend not to like vegetables that contain them. Children who are particularly reluctant to eat green vegetables may in fact be able to blame their genes, if they supertasters.

If avoiding the poisonous toxins in plants is such a good idea, why aren't we all supertasters, or at least tasters? In some populations over 90 per cent of people are tasters/supertasters and in others fewer than 40 per cent are sensitive, an observation that has led scientists to hypothesize that there may be a balance of risk and benefit of eating toxins that varies from place to place. What might be the benefit? It is possible that at suitably low doses some plant toxins may actually have a beneficial effect, for instance by acting as 'antioxidants' (see Chapter 4). One study that looked in detail at the proportion of non-tasters in different populations in Africa found that there was an association with endemic malaria: places with malaria had a higher proportion of non-tasters, so perhaps the bitter plant secondary compounds provide protection against malaria. This hypothesis is based on the fact that quinine, which was used to treat malaria from the 17th century until the 1940s, is a bitter plant secondary compound extracted from the bark of the cinchona tree. Quinine is thought to work by poisoning the malaria parasite.

Our genetic make-up probably also influences our sensitivity to other tastes. For instance, people are tasters, non-tasters, or supertasters for umami, and these have some association with genetic variation in the umami receptors. Likewise, there is some evidence that obese people tend to have a particular liking for sweet foods. However, as we shall see in the next section, our experience in childhood and later on can also have a dramatic effect on our food preferences.

Learning to like it or hate it

The American psychologist Paul Rozin estimates that about a quarter of the world's population eats chilli peppers, in spite of the fact that eating hot chilli is painful. In many countries there are hot chilli eating contests and there is even a world championship. The most challenging chilli pepper of all is the Dorset Naga variety that measures 1.5 million Scoville Heat Units, the scale in which hot chilli is scored. Rozin studied how we come to like eating such a painful food. He found that in a Mexican village, children aged between two and six years old were given gradually increasing amounts of hot chilli in their food, though not forced to eat it if they did not like it. By the age of five to eight the initial negative reaction to chilli had reversed and children voluntarily added hot sauce. A number of psychological influences may contribute to this remarkable reversal. These could include copying parents, peer pressure, and thrill seeking: eating hot chilli may be akin to bungee jumping or roller-coaster riding. At the same time hot chilli may elicit the production of endogenous opiates to turn the pain into pleasure, and by causing salivation might make dry food easier to chew.

The effect of early experience on food preferences starts in the uterus. For instance, in one study in France, babies born to mothers who had consumed anise flavour during pregnancy showed a preference for anise in the first few days after birth. During early postnatal life, the kind of milk a baby is fed has a substantial effect on its preferences for solid food. Human milk contains 14–31 mg/100 ml of glutamate, compared with 1.3 mg/ml in milk-based formula and a massive 440 mg/ml in hydrolysed protein-based formula. Babies fed on the hydrolysed protein formula, when compared with the others, have a stronger preference for cereals with a savoury, bitter, or sour flavour.

The interactions between taste and smell, referred to earlier, are also influenced by experience. Europeans and North Americans are much more sensitive to the aroma of almonds when a drop of sugar has been put on the tongue, but not when a drop of salt has been put on the tongue. However, Japanese people show the reverse effect. These results may suggest that our experience influences the way in which sensory inputs interact to create flavour. Europeans and North Americans normally experience the smell of almonds in sweet foods such as marzipan, while Japanese experience the smell of almonds in savoury foods such as pickled condiments.

Disgust

The old joke: 'What is worse than finding a slug in your salad?' Answer, 'Half a slug' is not about taste, it is about disgust. You feel sick, and might literally vomit, at the sight of it. Slugs are disgusting because of what they are rather than what they taste like, and furthermore, once the slug has been in contact with the lettuce, the lettuce itself becomes disgusting. But disgust is not inborn: any parent knows that toddlers go through a phase of sticking more or less anything in their mouth—slugs, faeces, soil, and so on. Furthermore, as I mentioned earlier, one culture's food delight may be another's disgust. In Britain, most adults would find eating lizards, insects, caterpillars, and dogs disgusting, but all of these are eaten in other cultures somewhere in the world. Disgust is probably a valuable defensive response in protecting us from eating foods that might be dangerous: most reactions of disgust are associated with animal foods, which are more likely than plants to contain dangerous bacteria.

Long-delay learning

There is a special form of learning that allows omnivorous animals to pick out dangerous foods and avoid them later on if they have had a bad effect, and is one of the ways in which our likes and dislikes are shaped by experience. It was first discovered in rats, by allowing them to consume sweetened water and then exposing them to radiation which made them sick some time later. The rats subsequently avoided sweetened water. The unusual features of long-delay learning are that it is based on a single experience and that the gap between the action (eating) and the consequence (sickness) may be several hours. In contrast, associative learning usually involves repeated exposure and a close link in time between behaviour and consequence (a few seconds). Pavlov's dogs learned to salivate at the sound of a bell after many trials, during which the association between the sound of the bell and the subsequent appearance of food was built up. They would not have learned the association at all had not the bell and the food been presented in rapid succession during training. Most of us have had the experience of getting violently sick some hours after a meal and subsequently associating sickness with the kind of food we had eaten, even if there was no direct causal link between the sickness and the food.

Religious taboos, cultural traditions, and evolutionary adaptations

Jews and Muslims do not eat pork, and Hindus avoid beef even though there are plenty of cattle in India. The anthropologist Marvin Harris has suggested that these, and many other religious or cultural food taboos, are not purely arbitrary, but have an ecological and economic rationale. He argues, for instance, that pigs are poorly adapted to the hot arid conditions of the Middle East where the Jewish and Muslim religions originated, and therefore it made ecological sense to enshrine the avoidance of pig meat, an unsustainable food, in Jewish and Muslim faith as a method of enforcement. Similarly, Hindus use cattle for many purposes: as beasts of burden, for their milk, and for their dung deployed as fuel and fertilizer. It was therefore an economically rational rule to protect cattle so that people did not, in effect, eat 'the goose that was laying the golden eggs'. In contemporary societies, patterns of behaviour that are good for the community as a whole, but not necessarily for the individual citizen, may be enforced by government regulations, such as restrictions on speeding and on smoking in public places, rather than by invoking the power of religion to ensure that everyone toes the line. It is implicit in Harris's hypothesis that without a religious or cultural taboo, individuals might be tempted to cheat, for instance by sneaking the odd pig into the village to live in the one shady spot and wallow in the one available mud pool. Harris applies similar ecological and economic logic to account for the distribution amongst cultures of numerous other kinds of dietary traditions and prohibitions such as eating horsemeat, dogs, humans, and insects.

Harris's hypothesis is plausible, but untestable, and his views have been much criticized by cultural anthropologists. Some anthropologists, the 'culture theorists', simply reject explanations of cultural traditions that are based on utilitarian hypotheses such as ecological or economic costs and benefits, while others have

I like it!

criticized Harris for making up 'Just So Stories' to fit the facts. Still another criticism is that, while he presents evolutionary hypotheses to account for different cultural food traditions, he does not try to explain how the traditions got started in the first place.

However, in one example, to which Harris refers, the links between evolution, culture, and food have been investigated in some detail: the evolution of lactase persistence.

Lactase persistence

About three-quarters of the world's adult population today is, to a greater or lesser degree, unable to digest lactose or milk sugar. If these adults drink milk, they experience symptoms that may include stomach cramps, bloating, diarrhoea, and gassiness. The frequency of lactose intolerance varies greatly from one place to another. In northern Europe, over 95 per cent of adults can digest milk sugar, whilst in parts of Asia fewer than 10 per cent are able to do so. The ability to digest lactose depends on the enzyme lactase that breaks down the milk sugar. Babies produce plenty of lactase to digest the lactose in their mothers' milk, but people who are lactose intolerant stop producing the enzyme after they are weaned. Hence the phenomenon of lactose tolerance is also called lactase persistence: the two go hand in hand.

In pre-agricultural human populations lactase persistence was very rare: analysis of ancient DNA from European skeletons about 7,000 years old shows that at that time Europeans were not lactose tolerant, and therefore that the lactase persistence genes have increased in frequency very rapidly in populations that have adopted livestock farming as a way of producing food. For these populations, milk from cows, sheep, goats, or other animals was a new and nutritious source of food and the ability to digest it would have been a great advantage, so there was rapid natural selection for the evolution of lactase persistence. So ecological factors, namely suitable habitat for keeping dairy animals, the cultural tradition of domesticating animals, and the genes for lactase persistence, have developed in an interrelated way during the past 7,000 years or so.

But why did our hunter-gatherer ancestors stop producing lactase after weaning? As I discussed in Chapter 1 when describing the consequences of cooking on our digestive system, 'there is no such thing as a free lunch' in evolution. Even the small energy cost of making the enzyme when it was not needed was sufficient to give an evolutionary advantage to stopping lactase production after weaning. It was only when it became advantageous to carry on making the enzyme into adulthood in order to benefit from a new source of food that the cost was worth paying.

Favism

Lactase persistence is the best understood, but not the only, example of coevolution of genes, culture, and food habits. About 400 million people in Africa, the Mediterranean, and the Middle East are unable to digest broad beans, a condition known as favism. This is unfortunate for these individuals, because broad beans happen to be one of the staple foods in these regions. The genetic mutation that causes favism alters an enzyme in the red blood cells, and this in turn reduces the ability of the parasite that causes malaria, *Plasmodium falciparum*, to survive in the human body. So the mutation is advantageous in areas where malaria is common, and the inability to digest broad beans seems to be an unfortunate by-product of selection for resistance to malaria.

All the genes in the body are represented twice in each cell, once on each of the relevant pair of chromosomes. Some individuals therefore will have one copy of the mutant gene and one copy of the normal version: they are what are known technically as heterozygotes. These lucky individuals may well have the best of both worlds, with some resistance to malaria and the ability to digest broad beans. It is only individuals with two copies of the 'favism gene' who are unable to digest the beans. These individuals may benefit from the fact that traditional recipes for preparing beans in areas with a high incidence of favism appear to make the beans easier to digest.

Both lactase persistence and favism are examples of genetic polymorphism, meaning that more than one variant of the relevant gene(s) coexist in equilibrium in a population. If one genetic variant confers an advantage on its owner, it might be expected that it would, over many generations, become very common and other variants very rare or disappear altogether. However, there are a number of reasons why this may not happen. One possibility, relevant to lactase persistence, is movement between populations. If the most favourable gene variant differs between populations, and if there is even a modest level of migration, the genetic polymorphism is maintained: so low levels of lactose intolerance have been maintained in European populations over many thousands of generations by immigration from areas with no tradition of drinking milk. As I have already hinted, perhaps the polymorphism of the favism gene is maintained because heterozygotes are at an advantage over either homozygotes, individuals with no copy of the mutant gene who are able to digest broad beans but are not resistant to malaria, or individuals with resistance to malaria but with favism.

Spices

When Alaric the Goth laid siege to Rome in AD 408 his ransom demand included gold, silver, silk, and one tonne of pepper. Pepper was immensely important in Roman cuisine. It was imported from India via the Red Sea, and thence across the desert to the Nile. Many hundreds of boats brought pepper from India each year, and one estimate of the value of a single cargo in the second century AD was that it would pay the annual salary of nearly 7,000 soldiers. Pepper continued to be an extraordinarily valuable commodity in medieval times. In England, the members of the Pepperers' Guild, founded in 1180, were the custodians of the King's weights, reflecting both the importance of accurate weighing of pepper and the importance of pepper as a barometer of commerce. In 1373, the Pepperers, having joined forces with the Spicers' Guild, renamed themselves the Company of Grossers, from which the word grocer is derived, because they were responsible for the accuracy of heavy weights or *peso grosso*. The Grossers soon took on an additional responsibility, for ensuring the purity of spices sold. This was called garbling, from the Arabic *gharbala*, meaning to sift or select.

All of this is to say that spices such as pepper have, for a very long time, been valuable and important elements of food. Spices were also important in stimulating international trade, exploration, and warfare, starting with ancient Egypt, Greece, and Rome and culminating in the colonial expansion and exploration by European countries in the 16th and 17th centuries, including Spain, Portugal, France, the Netherlands, and England.

Part of the reason for the high cost of spices in medieval times was their long journey by sea and land from Asia to Europe, via the horn of Africa, Egypt, and Venice. In 1492, the cartographer Martin Behaim listed twelve stages of their journey to the consumer. At each one levies and profits were added and he concluded: 'it need not be wondered that they are worth with us as much as gold'. It was against this background that the Portuguese King Manuel I funded Vasco da Gama's voyage in 1498 around the Cape of Good Hope to find a new sea route to the Indian Ocean, cutting out the middlemen, shifting the European centre of the spice trade from Venice to Lisbon, and dramatically reducing the cost of spices.

In the next few decades there followed a dispute between Portugal and Spain about the ownership of the Spice Islands, where many of the costly spices originated. These islands, the Moluccas, are now part of Indonesia. In 1519, the Spaniards discovered a western route, around the tip of South America, to the Indian Ocean, and both countries claimed the Spice Islands.

Earlier, in 1494, Portugal and Spain had signed the Treaty of Tordesillas, which essentially divided the known world outside Europe in two, by a meridian 370 leagues west of the Cape Verde Islands. All territory to the east of this (including what are now Brazil and Africa) was Portuguese, and that to the west, including the rest of South America, was Spanish. The treaty did not resolve whether the Spice Islands in the Pacific Ocean were to the west or east of the line. Following the discovery of both the western and eastern routes to the Pacific Ocean, a new treaty, the Treaty of Saragossa, was drawn up in 1529, in which the Portuguese purchased the rights to the islands for 350,000 gold ducats, because the Spanish cartographers, mendaciously as it turned out, persuaded the negotiating teams that the islands fell in the Spanish half of the world!

Why were, and indeed are, spices so highly prized in cooking? The answer to the question of why spices are used in cooking can be given in more than one way. On the one hand one might say that spices perk up bland food or cover up the taste of meat that has gone off. On the other it could be suggested that spices contain important micronutrients and other potentially beneficial ingredients such as antioxidants, or that they have antimicrobial properties that help to protect against food poisoning. The first two explanations are about possible immediate reasons for adding spices, the latter two are about the potential longer-term benefits that could, over time, have a survival advantage.

The American biologist Paul Sherman has analysed the use of forty-three different spices in 4,500 recipes representing the traditional cuisine of thirty-seven countries. The pattern he observed was striking and it fits with our intuitive expectation. More spices are used in cooking in hotter countries. But the pattern is subtler than this. Some spices, such as garlic, allspice, oregano, cinnamon, and cumin, have been shown to be especially powerful at killing bacteria in food, and these are the ones that are particularly favoured in hot countries. They are also more likely to be used in meat dishes that are prone to bacterial spoilage than in vegetarian dishes. The spices that are less effective in killing bacteria, such as ginger, aniseed, and celery seed, do not show such a clear association with hot climates. These patterns of spice use hold within as well as between countries. For instance, in China and the USA, both of which have large latitudinal gradients in climate, more spices are used in the hotter regions. But is it just a matter of using more spices where they grow in greater diversity and abundance? Not according to Sherman's analysis: there is no association between the number of spices grown in a country and average temperature. On the basis of his analysis, Sherman concludes that the extensive use of spices in cuisine has developed as a means of harnessing the antimicrobial properties of plants. This does not exclude the possibility that there are also nutritional or other benefits, but the antimicrobe hypothesis seems to be the most plausible explanation of variation in use of spices.

A crucial element of this story, and a possible criticism of its conclusions, is determining what counts as 'traditional cuisine'. Going back to the opening paragraph of this chapter, a visitor to Oxford from outer space would be hard-pressed to know what the traditional cuisine of England was. In fact, according to one UK-wide survey, a curry dish thought to have originated in Britain but based on the cuisine of the Indian subcontinent, chicken tikka masala, has overtaken fish and chips as the favourite dish of the British, and the then UK Foreign Secretary declared it to be a 'true British national dish'. The 'tradition' of English cuisine consisting of overcooked grey meat and soggy, flavourless vegetables is less than 200 years old. For instance, Robert May's *The Accomplisht Cook*, published in 1671, shows that at this time wealthy English households would flavour a roast joint of meat with spices such as ginger, coriander, fennel, cinnamon, cloves, saffron, and rose water.

It was only in the early 19th century, as a result of the Industrial Revolution and the emergence of a newly wealthy urban middle class, that a shortage of household cooks led to many untrained young women being employed as cooks, with the consequent decline in culinary standards for rich English households.

Global trade and cuisine: tomatoes in Italy

Cuisine has clearly changed in many parts of the world as a result of trade. The explorers who made contact with the New World brought tomatoes, potatoes, maize, quinoa, and avocados, as well as herbs such as coriander, back to Europe. Nowadays we associate chilli peppers and coriander with Asian cuisine, but in all probability these ingredients only arrived in Asia with the Portuguese explorers, who had encountered them in Latin America, in the 16th century. Before the arrival of the chilli, Asian cuisine used black or white pepper as a pungent spice. Chilli took over because it had a similar effect and was much easier to grow and therefore cheaper. The same is true of the other food plants that have been adopted outside their native country: if they were easy to grow, nutritious, or otherwise useful in cooking, they were often rapidly taken up.

However, the tomato in Italy is a remarkable counter example. Nowadays, Italian cuisine is often associated with the tomato: pasta al pomodoro, pizza Margherita, chicken cacciatore, and ragu alla Bolognese are just a few of the very familiar Italian dishes that include tomatoes. In the 1950s English cookery writer Elizabeth David criticized Italian cooking for containing too much tomato. But, remarkably, this is a tradition that dates back only as far as the 19th century. As the historian David Gentilcore has documented, there was a delay of more than 300 years between the arrival of the tomato in Italy in 1548, from the New World, via Spain, and its widespread use in cooking. The Italians adopted other New World crops such as maize, beans, chillis, and tobacco, before they accepted tomatoes.

Gentilcore suggests a number of possible reasons for this. The early tomatoes were sour because sweetness had not yet been bred into them. They were identified by botanists as belonging to the Solanaceae, the same family of plants as deadly nightshade (along with aubergines, potatoes, and tobacco) and were therefore treated with suspicion. One authority described them as 'dangerous and harmful' and their smell alone was said to cause 'eye diseases and headaches'. Tomatoes were nevertheless grown as ornamental plants. The Italian name pomo d'oro (yellow fruit), as opposed to the Aztec 'tomatl', may have been adopted as an allusion to the tree that bears golden apples in Greek mythology.

The tomato's acceptance as a food was gradual: in the mid 17th century there are references to its use as a condiment, to make a salsa with chilli peppers, and by the late 18th century recipes in cookery books were appearing. Methods of preserving tomatoes for use in the winter were also being developed. These included drying halved tomatoes (equivalent to modern sun-dried tomatoes) and making a paste from boiled tomato purée that was then dried in sheets in the sun. But dishes such as pasta with tomato sauce and pizza Margherita were probably 19th-century inventions, the latter formally named after the queen of Italy in 1889. The crucial development of canned tomatoes and tomato purée took place in Britain and the USA: in fact, the emigrant Italians influenced the development of tomato-based cuisine in the USA and the breeding of new varieties, which were later reimported into Italy.

In this chapter we have seen how our food preferences are affected by many influences, including our sensory receptors and brain, our genetic make-up, our infant and childhood exposure, as well as our social and cultural environment. As omnivores we have evolved to be both cautious about the potential risks associated with unfamiliar foods and flexible enough to embrace new food experiences. Food for us humans is more than fuel. It is part of our culture, our social life, our religions, and our pleasure.

But some worry about the dangers that might lurk on their plate. In one study that compared French and American attitudes to food, when asked what they associated with fried eggs, the French thought of breakfast, the Americans of cholesterol. In the next chapter I will examine some concerns about food, and ask whether or not they are justified.

Chapter 3
When food goes wrong

Introduction

20 March 1996 was a pivotal day in the modern history of food safety. It was the day on which the UK Secretary of State for Health, Stephen Dorrell, told Parliament that ten young people had contracted a fatal and incurable new disease, called new variant Creutzfeldt-Jakob Disease or vCJD for short. He said that these victims had probably caught the disease as a result of eating beef or beef products from cows suffering from bovine spongiform encephalopathy (BSE), commonly known as Mad Cow disease. BSE was a recently diagnosed fatal disease of cattle, with early symptoms of unsteady walking, leading eventually to failure of the nervous system. In the late 1980s, an expert committee had examined the possible risks to humans and concluded that they were 'remote'. In 1992, the then Agriculture Minister John Gummer had posed for a photo shoot with his young daughter eating a hamburger, with the quote 'British beef is perfectly safe'. This turned out to be far from the truth. The BSE crisis had a dramatic effect on international trade in beef and cattle, on confidence in scientists, politicians, and the governance of food safety. Although it was primarily a UK problem, it later emerged that BSE had affected cattle, and humans, in other countries in Europe, North America, and Asia: by the early 2000s it had

become a global issue, with trade embargoes and other restrictions affecting many countries.

Mad Cow disease

BSE and vCJD, otherwise known as human BSE, belong to a family of mammalian brain diseases, the transmissable spongiform encephalopathies, or TSEs for short, that are caused in a most unusual, perhaps unique, way. Almost all infectious diseases are caused by microorganisms such as bacteria, viruses, and fungi. These microorganisms have their own genetic code and reproduce inside the body of their host. Viruses are the simplest, in that they consist of nothing but genetic material (DNA or RNA) in a protective protein coat. Inside a host cell, they hijack the cell's biochemical machinery to make copies of themselves. TSEs are caused by something even simpler than a virus: a protein called a prion protein, with no genetic material. We all have prion proteins in our brains, and they play a crucial role in normal functioning of brain cell membranes. All proteins, which are the building blocks of body tissues, are made up of a string of small molecules called amino acids, folded into characteristic, complex 3-D shapes, rather like a scrunched-up length of wire. TSEs are caused when the brain is infected with a copy of a prion protein that is folded in an abnormal way. This abnormal, or mutant, protein acts as a template, causing the body's own prions to fold abnormally, until they build up inside nerve cells in the brain into a 'plaque' or tangle of protein that kills the cell. The dead cells appear as tiny holes in the brain, creating a sponge-like appearance when thin slices of brain tissue are viewed under a microscope; hence the name spongiform encephalopathy. The abnormal prion is unusually resistant to being destroyed by heat and by the enzymes called proteases that normally break up proteins. TSEs had been known about for many decades before BSE, and the hypothesis that they are caused by prions dates back to the 1960s. In 1997, the American biologist Stanley Prusiner won a Nobel Prize for his discoveries showing that prions are the infective agents in TSEs.

Another American scientist, Carlton Gajdusek, had won a Nobel Prize, in 1976, for the first demonstration of a TSE in humans. He studied the South Fore tribe of New Guinea and discovered that their habit of ritual cannibalism of dead relatives transmitted a very unusual and invariably fatal brain disease called kuru. The symptoms, as with BSE and vCJD, included gradual loss of muscle control and eventually loss of control of all bodily functions. Kuru has a very long incubation period, on average about fourteen years, although for some people it is much longer than this, because individuals differ in their genetic resistance to the disease. It is thought that kuru may have originated in the South Fore tribe through one individual who developed the disease at the start of the 20th century, and when he or she died, other members of the tribe ate the infected brain. Kuru was more common in the South Fore tribe among women and children than in men, because in the cannibalistic ceremonies, men ate the choice cuts of meat while women and children ate the brains and internal organs of their dead relatives, where most of the infective agent was concentrated.

Mad Cow disease is also thought to have spread through the UK cattle population as a result of cows eating infected dead relatives. The practice of feeding dairy and beef cattle with a protein supplement called meat and bone meal (MBM)—ground-up remains of slaughtered cattle and sheep that had been sterilized by heating to a high temperature—was not new. But in the 1980s the animal feed industry in the UK lowered the sterilization temperature to save costs, and this may have allowed the heat-resistant mutant prion to survive and infect cattle. The disease could have started as a result of a spontaneous new mutation in the genes that code for the production of the normal prion in cattle, or it could have been a mutation of another TSE, called scrapie, which is not a risk to human health, that had been known in UK sheep since the early 18th century.

The custom of feeding meat and bone meal to cattle was widespread in many parts of the world, so why did the UK become the world centre of BSE? It seems likely that three factors contributed. First, the change in the sterilization temperature for making meat and bone meal referred to above. Second, in the 1970s UK farmers started to feed meat and bone meal to young dairy calves and these animals may have been more susceptible than adult cows to the disease. Third, the UK had a very large sheep population, with about 5,000–10,000 cases of scrapie per year. If, as mentioned above, scrapie was the source of the new disease, this might explain why the UK was particularly at risk. Between the late 1980s and early 2000s, over 180,000 cattle in the UK were slaughtered and incinerated because they were identified as infected with BSE, and possibly more than one million infected animals went into the UK food chain. Once the source of spread of the disease had been identified, feeding MBM was banned in many countries, cutting off the source of new infections, and the disease rapidly declined and had more or less disappeared by the end of the first decade of the 21st century.

Human BSE, or vCJD, was recognizable by examining the brains of people who had died. It had all the features that instil fear and dread. It was a new and mysterious disease, it affected young people, it was incurable, it had dreadful symptoms of slow decline into a vegetative state before death, and it had a long, unknown incubation period that could, if kuru was anything to go by, be measured in decades. No one knew how many people would die, but experts estimated, from the amount of infected material that went into the food chain, and assessment of the likely infective dose, that the number of deaths in the UK could be between tens of thousands and hundreds of thousands over twenty or thirty years. By a stroke of luck, these estimates were wildly pessimistic. In the UK fewer than 200 people died and in the rest of the world about thirty people died. The principal reason for the difference between the prediction and the outcome is that humans are much more resistant to the disease than had been assumed by extrapolating from experiments with mice, and therefore the doses of prion that people ate were not sufficient to cause an infection to spread.

But in the late 1990s and early 2000s there was widespread public concern in many countries around the world, although there were far fewer cases of BSE outside the UK. For instance, in Canada nineteen, and in the USA four, cases had been confirmed up to 2012. By the mid 1990s, measures had been put in place in the UK, followed later by other countries, to prevent infected material getting into the human food chain. These included banning the consumption of the parts of the animal, referred to as 'specified risk material', such as the brain, spinal cord, and some internal organs, that would contain most of the abnormal prion if the cow were infected, and allowing only young cattle, under the age of thirty months, to enter the food chain. The reasoning behind this was that young animals, even if they had the disease, would not have significant amounts of infective prion in their bodies, because the disease develops very slowly in cattle over five years. In the absence of a test for BSE, which was not developed until the early 2000s, this was a reasonable protection measure. Once a reliable test had been developed, this was used to exclude infected cattle from the food chain.

Could there ever be another food crisis like BSE? The answer has to be yes, because animal diseases evolve in unpredictable ways, and can sometimes become a risk to humans. No one can guarantee that food is perfectly safe, even though this is what politicians like to claim. The UK government commissioned a very long and very expensive enquiry into BSE, conducted by one of the most senior judges in the country, Lord Phillips of Worth Matravers. He concluded that: 'the Government does not set out to achieve zero risk, but to reduce risk to a level which should be acceptable to a reasonable consumer'. In the rest of this chapter we will look at examples from the past and the present to see what risks lie in food.

Milk

In the 21st century in rich countries, consumers tend to be risk averse, and expect the food that they buy to be safe. It is remarkable how what is seen as an acceptable risk has changed, even in one lifetime. When people buy a plastic jug, cardboard carton, or glass bottle of milk, they do not think of it as a potential source of danger, yet in the 1930s in the UK, an estimated 2,500 deaths per year resulted from drinking milk infected with the bacterium that causes bovine tuberculosis. This risk would have been easy to control, by the simple expedient of requiring milk to be pasteurized and cattle to be tested for TB. However, the UK Parliament refused to pass legislation to mandate pasteurization of milk for public sale until 1949, on the grounds that individuals should be free to drink raw milk if they wished, and that the risk was not great enough to warrant an intrusion into this freedom. Today there would be a national scandal in any rich nation if milk were as dangerous as it was only a few decades ago.

But milk in the 1930s was nevertheless much safer than it had been a few generations earlier. The 18th-century English writer Tobias Smollett was very interested in food, and in his picaresque novel *The Expedition of Humphry Clinker*, there are colourful, comic, descriptions of food in different parts of Europe. Here is what he says about milk in London in the good old days:

> Milk itself should not pass unanalysed, the produce of faded cabbage and sour draff, lowered with hot water, frothed with bruised snails, carried through the streets in open pails, exposed to the foul rinsings discharged from doors and windows, spittle, snot and tobacco-quids from foot-passengers, overflowings from mud-carts, spatterings from coach wheels, dirt and trash chucked into it by roguish boys for the joke's sake, the spewings of infants who have flabbered in the tin measure which is thrown back in the condition among the milk for the benefit of the next customer; and, finally, the vermin that drops from the rags of the nasty drab that vends this precious mixture, under the respectable denomination of milk-maid.

Smollett employs a degree of poetic licence, but milk in the 18th and first half of the 19th century was not too far from his caricature. Milk in London was routinely watered down, thickened with flour, sweetened with carrot juice, and coloured with a yellow dye. In the Paris siege of 1870, it was reported that infant mortality fell by 40 per cent because mothers breastfed their babies rather than giving them dangerous cows' milk. One of the most prominent food scandals of the mid 19th century in the USA was the 'swill milk' scandal. Milk was produced in New York in unspeakable conditions by cows fed on the hot grain mash left over from the brewing industry. The *New York Times* of 22 January 1853, in an article entitled 'Death in the Jug', described how this milk was then adulterated:

> To every quart of milk a pint of water is added; then a quantity of chalk, or Plaster of Paris to take away the bluish appearance produced by dilution; magnesia, flour and starch to give it consistency, after which a small quantity of molasses is poured in to produce the rich yellow colour which good milk generally possesses. It is now fit for nurseries, tea tables, and ice-cream saloons, and is distributed over the City, insidious, fatal and revolting poison.

Beer: the purity law

While milk in many places was often an unsavoury, if not dangerous, drink, beer in central Europe was closely regulated. The first national food standards law was the 'Reinheitsgebot' or purity law introduced in Bavaria in 1516. According to the original law, beer could be made from only three ingredients: water, barley, and hops. Yeast had not yet been discovered and the fermentation was brought about by naturally occurring yeast on the barley. Remarkably, the yeast used in making lager beer came originally from Patagonia: how it made the journey to Bavaria is a mystery. The reason for the law was partly to do with safety: beer was sometimes flavoured with bitter plants other than hops, including some that were poisonous. It was also partly to ensure that wheat, essential for bread making and often in short supply, was not diverted into beer production. Although the purity law is no longer a legal requirement, many German beers still follow the rules and proclaim it as part of their brand.

Food adulteration

In the 19th century it was not just milk that was routinely adulterated. Toxic chemicals and other additives were often used in other day-to-day staples. Sweets were coloured with lead, mercury, or copper salts, pickles made green by adding copper, vinegar sharpened with sulphuric acid, tea leaves were mixed with dyed blackthorn leaves, black pepper with dirt, white bread flour with crystalline aluminium sulphate made by adding urine to shale containing aluminium salts, and so on. Adulteration of food goes a long way back in history: there are accounts of adulteration in Roman times and it still goes on today. In the 1980s, certain Austrian wine producers were caught adding ethylene glycol to their wine to sweeten it, and as we shall see below, much more recently milk in China was adulterated with fatal consequences. The scale of the problem in the early 19th century was highlighted in 1820 when Frederick Accum, a German-born chemist living in London, published his *Treatise on Adulterations of Food, and Culinary Poisons*, in which he laid bare the risks and frauds of adulteration. But the authorities in Britain and the USA were reluctant to act because of the prevailing free market philosophy, and it was many decades later that legislation and enforcement of food standards became the norm. When Accum pointed out that tea was routinely adulterated, *The Times* of London blamed this on high import taxes that encouraged criminal activity!

Today, as a result of regulations and enforcement, food standards in rich countries are incomparably higher than they were in the mid 19th century. But this does not mean food is risk free. In the first decade or so of the 21st century there were notable food contamination and food poisoning incidents. Chinese powdered milk and infant formula were adulterated with melamine to increase their apparent protein content, because the standard test for protein also reacted with melamine. The incident affected at least 300,000 people and killed at least six babies. In 2011, a major outbreak of food-borne illness caused by a particularly dangerous strain of the bacterium *E. coli* in Germany caused even more fatalities, killing about fifty people and affecting 4,000. The infection caused a serious, and sometimes fatal, kidney disease called haemolytical uremic syndrome. The source of the outbreak was difficult to pin down. At first the German authorities blamed Spanish cucumbers, which led to a BSE-like ban on Spanish salad vegetables by some countries. There was also an eerily familiar replay of John Gummer in the early 1990s when Rosa Aguilar, the Spanish Agriculture Minister, appeared on television with a cucumber in her mouth saying, 'Spanish cucumbers are perfectly safe'! In the end the outbreak was thought to have stemmed from bean or fenugreek sprouts, possibly grown on an organic farm in north Germany.

Chemicals in food: natural and artificial

In consumer surveys of worries about food safety, people frequently voice concerns about 'chemicals in their food'. One international food outlet claims to 'shun obscure chemicals' in its food. This is a pretty staggering claim, since food is nothing but chemicals, some of them quite obscure.

Of course, the worries are about the chemicals that are added during production or processing of food, not the chemicals that are there 'naturally'. This is because of a fallacious belief that 'natural is good, artificial is bad'. However, many of the plants we eat regularly contain small amounts of natural poisons or toxins that could cause cancer. As we saw in Chapters 1 and 2, these are the plant's natural defences against insects, fungi, and other enemies. The American toxicologist Bruce Ames in a famous paper entitled 'Dietary pesticides (99.99 per cent all natural)', reported that 99.99 per cent of the 'pesticides' in the American diet are naturally produced defences of plants and that about half of these natural chemicals are carcinogens at high doses in rodents. Ames's findings have been summarized by alluding to the fact that a single cup of coffee contains natural carcinogens equivalent, in toxicity terms, to a year's worth of carcinogens from pesticide residues consumed in food.

But, and this is an important but, these results do not mean that people should stop drinking coffee or eating vegetables and fruit. An early 16th-century Swiss physician called Philippus Aureolus Theophrastus Bombastus von Hohenheim, also known as Paracelsus, is generally recognized as the father of the science of toxicology. His key contribution was summarized in a sentence: 'All things are poison, and nothing is without poison; only the dose permits something not to be poisonous.' In other words, something that is harmless at a low dose may be dangerous if a lot of it is consumed, and vice versa. Even pure water, if drunk to excess, can be fatal, and although there may be many potential carcinogens in food, both natural and synthetic, whether or not they present a real danger depends on the dose.

Just to illustrate the point, oranges are sometimes treated after harvesting with a pesticide called imazalil, to prevent mould forming on the peel. Imazalil could in theory be a carcinogen. Based on experiments on rats, a potentially dangerous dose for humans would require eating over 12,000 oranges including the peel. But oranges also contain about 70 mg of vitamin C per fruit, close to the recommended daily intake of this essential micronutrient. But as with imazalil, too much vitamin C is harmful. Extrapolating from rats, the number of oranges that could deliver a fatal dose of vitamin C is about 8,000. In short, vitamin C poses a bigger risk than imazalil.

In many, if not most, countries there are strict regulations governing the levels of pesticides and other synthetic chemicals, such as colours and flavours, in food. Safety is assessed by experts, using evidence from experimental tests and other information such as the chemical properties and structure of the substance. The starting point is often experimental data on the toxicity of the chemical, or chemically related substances, in laboratory tests on animals such as rats or mice. The critical measurement is the 'dose–response relationship': an indication of adverse responses, ranging from short-term effects such as vomiting to long-term responses such as the development of cancers, to different doses of the substance. From this it is possible to estimate a 'no observable effect' level of the substance. In transferring this to humans, a safety margin is built in, for instance by taking the 'safe level' as 100th or 1,000th of the 'no observable effect' level in rodents, standardized for body mass.

Those who worry that this kind of process is not robust enough to detect potential problems tend to argue that extrapolating from rodents to humans, even with built-in safety margins, is not justified, or that cocktails of different synthetic chemicals in food may have unexpected effects not predicted from testing them one at a time, or that there may be very long-term effects not picked up by short-term toxicological tests. These are perfectly reasonable concerns and the science of toxicological risk assessment often involves informed judgements. However, the fact remains that synthetic chemicals such as pesticides, flavours, and colours are much more rigorously scrutinized for possible risks than are the naturally occurring toxins. Potatoes, for instance, would probably be banned if they were subject to the same scrutiny because they contain poisons called glycoalkaloids at levels far higher than 100th of the 'no observable effect' level. The generally accepted safety level is 200 mg/kg fresh weight of potatoes, and the average level in commercial potatoes is typically about half of this.

One moral of this section is that whenever there is a newspaper story about 'cancer scare in baby food' or 'grilled meats can cause cancer', one should ask whether or not the dose consumed in a typical diet is actually enough to cause harm. Another is that synthetic chemicals in food are not necessarily a bigger risk than natural chemicals. A third is that one cannot avoid eating foods that contain potential toxins.

The dictum that 'it all depends on the dose' is not the same as saying that 'more is always worse'. As the example of vitamin C and oranges illustrates, many chemicals in our diet have positive benefits, or may be essential for survival, at low doses, but are poisonous if eaten in excess: consuming a small amount of vitamin C is better than consuming none, but consuming a very large amount could be dangerous. We will look at this in more detail in Chapter 4.

Acrylamide

In 2002, Swedish scientists discovered that many plant-based starchy foods that have been cooked at high temperatures, for instance French fries, potato crisps (chips), and bread, contain a chemical called acrylamide. Acrylamide, formed as a result of a chemical reaction between glucose and the amino acid asparagine, can cause cancer in rodents, and the toxicological advice is that people should eat as little as reasonably practical. But none of the national or international food safety regulators advised people to stop eating the many foods that contain acrylamide. It is not a new risk, since humans have been cooking food for several hundred thousand years (see Chapter 1), and it is likely that the levels at which most of us consume acrylamide is not going to add measurably to our risk of cancer. Several studies, from Europe and the USA, have not shown any links between acrylamide intake and increased cancer risk. However, this kind of study may be difficult to interpret because the data rely on people's accurate recollection of what they have eaten, and also there are many other factors that influence a person's cancer risk (see also Chapter 4).

Hyperactivity in children, and food additives

Just how difficult it can be to decide whether or not artificial chemicals in food cause harm is highlighted by the so-called 'Southampton study'. There is a long-standing debate about whether or not food colours such as tartrazine, sunset yellow, and carmoisine, contribute to attention-deficit hyperactivity disorder (ADHD) in children. Many parents try to avoid giving their children soft drinks or sweets with these colours in them. Parents may be convinced that their children are overactive or less well behaved after drinking brightly coloured fizzy drinks, but there are several problems with this as a reliable source of data. First, parents may be biased in their observation by their prior conviction that there is an effect; second, the fizzy drink might give the child an energy boost, hence making the child more active, and this could be mistaken for an effect of the colouring agent; and third, fizzy drinks may be drunk in situations such as parties where the occasion could contribute to children being very active.

The Southampton study tried to nail the question once and for all by carrying out a randomized, double-blind, placebo-controlled trial. In other words, children were assigned at random to different treatments. No one involved in administering or recording knew which treatment was given to which child, and children were given either a drink with the artificial colours under investigation or an indistinguishable placebo. The study tested two age groups, three-year-olds and eight- to nine-year-olds, with two mixes of colours. The authors concluded that their results showed an effect of the food additives on hyperactivity in children, as assessed by teachers and parents. But the European Food Safety Authority took the view that the study did not 'substantiate a causal link between these individual colours and possible behavioural effects'. There may be a number of reasons for this conclusion, including the fact that the study did not look at individual colours, but only at mixtures of four colours combined with the preservative sodium benzoate. Furthermore, only one of the two mixtures had an effect on hyperactivity, measured by a composite score of different ways of assessing the children's behaviour. So even after a 'gold standard' study, the matter of whether or not food colours cause hyperactivity is unresolved.

Food allergy

In the UK, about ten people per year die as a result of a food allergy. Most die from eating peanuts or tree nuts, the remainder from allergy to dairy products or fish/shellfish. In population surveys, about one-third of respondents say that they are allergic to one kind of food or another. But the real prevalence of allergy is much lower than this, affecting about 5 to 7 per cent of children and 1 to 2 per cent of adults in the UK. People who think they suffer from food allergy may in fact have a food intolerance, for instance to wheat or milk, rather than an allergy (see Chapter 2 for a discussion of milk intolerance), or simply a strong dislike of a particular food. True allergy, as we shall see, depends on the immune system and can mean that even minute doses of a food can be very dangerous, whilst intolerance is not dependent on the immune system and the response is much more linked to the dose. The prevalence of food allergy in the UK is increasing: peanut allergy, for instance, more than doubled in the first five years of the 21st century, and furthermore, new allergies to foods such as kiwi fruit, sesame seeds, and gooseberries are appearing. No one knows why food allergies are on the increase; it might be linked to the general increase in other so-called atopic allergies, such as asthma, hay fever, and eczema. Allergy in general is on the increase in most countries in western Europe as well as in the USA, Canada, Australia, and New Zealand. The increase is mainly seen in wealthy countries.

Before trying to explain the possible causes of the increase in allergies in general, and food allergies in particular, we need to stand back and look briefly at what an allergic reaction is. The following is a greatly simplified account.

An allergic reaction to food, or other stimuli such as dust, pollen, bee stings, and soap, is the result of the body's immune system directing its defences at the 'wrong' stimulus. The immune system, which has many components, is the body's mechanism for defending itself against invading organisms that could cause harm, such as bacteria, viruses, and larger parasites such as worms and flukes. An immune response usually includes a number of stages: recognition of a foreign object, inflammation of the affected area, and destruction of the invaders either by chemical attack or by special cells that devour it. Food allergy arises when the immune system 'thinks' that a protein in food is an invading parasite and sets in motion the response that would normally destroy the invader. Food allergies, along with so-called 'atopic allergies' such as asthma and eczema, are often associated with a particular kind of recognition system, or antibody, called immunoglobulin E (IgE).

Antibodies are the first line of defence in the immune system: they attach to the foreign protein and act as a signal for other parts of the immune system to destroy the invader. Once IgE is attached to the foreign protein, the immune system starts to release chemical messengers such as histamine that cause blood vessels to dilate, which results in the swelling or redness of the skin that is often associated with infection. Dilation of the blood vessels allows more blood to flow into the affected area, carrying components of the immune system, which destroy the invader. The inflammation associated with the IgE immune response can be very dangerous. It makes it difficult for asthmatics to breathe; and in people with food allergy it can cause 'anaphylaxis', which is often fatal if not treated immediately. Anaphylaxis is the immune system going into overdrive, so that the swelling reaction, rather than being localized, affects large parts of the body including the throat.

Now we can return to the question of why allergy in general, and food allergy in particular, is increasing in many countries. It is unlikely to be a result of any genetic change, because the increase in prevalence has occurred over a very short timescale of a few decades.

One idea, first put forward in the 1980s, is called the 'hygiene hypothesis'. The essence is that when we grow up in conditions that are too hygienic, our immune system is underemployed and therefore attacks inappropriate 'invaders' such as the proteins in food. There are several lines of evidence that support this hypothesis. First, allergy is much less prevalent in developing countries where hygiene standards are lower and children are exposed to many infections when they are young. Second, some studies have indicated that children in larger families in traditional rural communities are exposed to more infections and are less likely to develop allergies. Third, in Germany, before reunification, children in the East were less likely to suffer from allergy than their Western counterparts, but this difference disappeared after Germany was reunified in 1989 and the East adopted Western hygiene standards associated with greater affluence. In Finland, one study found an association among teenagers between the prevalence of atopic allergy and environmental biodiversity, including bacterial diversity: children living in less biodiverse areas were more likely to have allergies. The hygiene hypothesis is sometimes summarized as 'a little bit of dirt does you no harm'. However, this does not mean that it is a good idea to eat dangerous organisms that cause food poisoning!

Another idea, which is not incompatible with the hygiene hypothesis, is that early exposure to a food could lead to priming of the immune system to react against the proteins in this kind of food later on in life. The UK Department of Health used to advise pregnant women with a history of any kind of allergy in the family against eating peanuts during pregnancy or during breastfeeding. However, some studies have indicated that exposure to peanuts in early life can actually *reduce* later allergy, so the advice has been withdrawn.

Food allergies also interact with other allergies: for instance, people who become allergic to birch tree pollen may, as a result, also become allergic to kiwi fruit. There is no doubt that repeated exposure to an allergen increases the immune system's response, so that allergy builds up over time, but the way in which exposure in early life interacts with immune response in the hygiene hypothesis is still not clear. Equally, it is not clear why children have a higher prevalence of allergy than do adults.

One major international study revealed a correlation between diet and allergies such as asthma and eczema among children: individuals eating more fast food were more prone to allergy, while those eating more fruit and vegetables were less prone. However, this does not show that allergies are caused by eating junk food.

Food poisoning

Almost everyone has suffered from a bout of 'food poisoning', or more accurately from a food-borne illness. Worldwide, untreated food- or water-borne illnesses are a major cause of infant mortality, accounting for an estimated 500,000 deaths per year, often through dehydration that results from extended periods of diarrhoea. Some bouts of food poisoning last for only 24–48 hours, during which the sufferer may have severe diarrhoea and/or vomiting. This kind of food poisoning is usually caused by a group of viruses known as 'small round structured viruses' or 'rotoviruses'. The virus gets onto the food during preparation or eating, usually from someone's hands. The viruses are very contagious, so can easily spread within a group such as a family unless people are scrupulous about hand-washing.

Bacterial food poisoning is caused by pathogens such as *Salmonella*, *Campylobacter*, *Escherichia coli*, *Listeria*, and *Clostridium*. It is usually much more serious than viral food poisoning. In the UK, for instance, it is estimated that around

500 people a year die of bacterial food-borne illness, and in major incidents, such as the *E. coli* outbreak in northern Germany in 2011, already referred to, many tens of people may die. The bacteria that cause these more serious forms of food poisoning often live in the raw materials that go into food. For instance, *Salmonella* and *Campylobacter* are found in chickens, which is why eating raw or undercooked poultry is risky. They can also be transferred from raw meat to other food in the kitchen. In the USA the Centers for Disease Control and Prevention (CDC) record outbreaks of bacterial food poisoning. In 2012, sixteen investigations were listed, including outbreaks from raw salads, fruit, ground beef, and poultry. Overall, the CDC estimates, food-borne disease in the USA causes 5,000 deaths, 325,000 hospitalizations, and 76,000,000 illnesses.

Good and bad bacteria

By no means all bacteria are harmful to humans. In fact, the average human being probably contains about ten times as many bacterial cells as human cells (a hundred trillion versus ten trillion: because bacterial cells are much smaller than human cells these ten trillion weigh about 1 kilogram). Most of these bacteria live in our digestive tract and some have positive effects. It has been estimated that roughly 10 per cent of our energy intake is a result of bacteria helping us to digest parts of food that we would otherwise be unable to utilize. For instance, mothers' milk contains carbohydrates called glycans, which are only digestible because certain bacteria in the gut convert them into sugars. Other bacteria in the digestive tract may be involved in causing certain chronic diseases such as stomach and bowel cancer as well as cardiovascular disease, although the details are still not understood. It is possible to buy foods called probiotics, especially yoghurt that contains the bacteria that ferment milk (see Chapter 1). It is claimed in advertisements that these probiotics increase the populations of 'good bacteria' in the digestive tract. There is no evidence that probiotics are of benefit to healthy individuals, but some randomized control trials have shown that they may help to alleviate some kinds of infant diarrhoea. For other bowel diseases the evidence is equivocal.

Organic food and genetically modified food

Consumer surveys show that the main reason people buy organic food is because they believe it to be safer or more nutritious. If there were a difference in safety, this would present a problem for food retailers, because it would mean that 97 per cent of the food they sell, which is conventionally produced, is in some way 'less safe' than the 3 per cent that is organic.

The organic movement has its origins in the writings of a late 19th-century Austrian anthroposophist (student of the non-material, spiritual world that is not accessible to science) called Rudolf Steiner. During the first half of the 20th century Steiner's ideas were modified and incorporated into the foundations of the contemporary organic farming movement, especially by Lady Eve Balfour and Sir Albert Howard in the UK and J. I. Rodale, who coined the word 'organic', in the USA. The essence of their philosophy is that food grown in a more 'natural' way, with manure and crop rotation instead of chemical fertilizers, and with natural pesticides rather than artificial chemicals, is both more wholesome and more sustainable. Different organic certification bodies have different criteria for organic farmers, but on the whole they all permit fewer pesticides, drugs for preventing animal diseases, and synthetic fertilizers, than are used in conventional modern agriculture. In this regard, organic farming has more affinity to 19th-century farming or to subsistence farming in poor countries, than to late 20th- or 21st-century farming in the developed world.

Organic food production uses fewer pesticides but is not pesticide free: the approved list for the Soil Association, the UK's largest trade and accreditation body for organic food, includes substances such as copper sulphate, rotenone, and sulphur because they are regarded as natural. Does the use of fewer pesticides and other synthetic chemicals in food production and manufacture make organic food safer?

We have already seen that natural toxins are not necessarily any safer than artificial or synthetic chemicals, and that the toxicity of most substances depends on the dose. The synthetic drugs administered to farm animals to reduce their burden of disease may help to reduce the prevalence of pathogens that could be a danger to human consumers. This could in part explain why a Danish study found that organic chickens were three times more likely to contain *Campylobacter* than were conventionally reared chickens. However, other studies in America have found no differences in the frequency of finding *Salmonella* or *Campylobacter* in chicken in the shops.

Conventionally produced crops, fruit, and vegetables are often treated, pre or post harvest, with pesticides to prevent fungal growth, while their organic equivalents are not. In theory, this could mean that organic foods are more dangerous because fungi often produce harmful toxins, such as afflotoxin in peanuts and ergot in wheat, which used to kill many thousands of people in the Middle Ages in Europe. In short, there are possible reasons for worrying about the safety of organic, as well as conventionally produced food. However, the bottom line is that no studies have shown consistent differences in safety between organic and conventional food. Consumers of organic food, however, could argue that there may be unknown long-term adverse effects of agrochemicals, or that cocktails of different chemicals may have unpredictable effects.

The organic movement is opposed to genetically modified (GM) food. In some ways this is a paradox, since one aim of GM technology is to provide plants with built-in insecticides, so reducing the need to treat crops with chemical sprays. Organic farmers are allowed to spray their crops with an insecticide made of the bacterium *Bacillus thuringensis*, but are not allowed to grow crops with the 'insecticide gene' from the bacterium introduced into the crop plant.

There are essentially three main objections to GM foods. First, they are 'unnatural'; second, they pose a threat to the environment; and third, they are a risk to human health. Here we are concerned only with the third issue: I will return to the others in Chapter 5. How do we know whether or not GM foods are a risk to human health? Each new GM variety is assessed for its safety before it is approved for consumption, using principles similar to those for toxicological risk assessment. The experts consider whether it might be allergenic, toxic, or have some other adverse effect on consumers. One important element of this risk assessment is called 'substantial equivalence'. Because genetic modification of plants is a precision technology, scientists can specify the chemical changes in the plant that will result from a particular modification. If the experts conclude that a new genetically modified food or crop is substantially equivalent, it does not differ from conventionally produced food in the chemical properties that could affect toxicity or allergenicity. It therefore does not pose a new risk. Sceptics say that this is not the same as testing the GM foods on humans over many years and they point to some studies on laboratory rats that appear to show adverse effects of diets containing GM products, although these studies have been shown to be inadequate in many ways. In practice, populations such as those of the USA have been consuming GM ingredients such as soya and maize for well over a dozen years without any detected ill effects, although such effects might be very difficult to tease out.

Conclusion

In this chapter I have reviewed some of the risks that might lie in the food we eat. I have also suggested that our subjective perception of risks may differ from the objective scientific evidence. What are the biggest risks in food today? Pesticide or pollutant residues? Food-borne illness? Allergens? Additives such as artificial colours or flavours? These are among the worries that are typically chosen in opinion polls when surveys offer a prompt list from which to choose. One approximate way of comparing risks is to count the number of food-related deaths attributed to different causes. This is oversimplified because it is not always easy to determine a single cause of death, and it ignores morbidity. Nevertheless, it gives an indication of the relative size of risks. For the UK, about 500 deaths per year are caused by food-borne illness and in the region of ten by food allergy. No deaths are attributed to pesticide poisoning, GM food, or additives. But the really big risks associated with our food are the contribution that our diet makes to the risk of the major chronic diseases that between them will kill more of us than anything else: cancer, cardiovascular disease, and stroke. Whilst it is difficult to assign an exact number, the dietary contribution to death from these causes may be equivalent to in the region of 100,000 deaths per year. In the next chapter we will discover where this estimate comes from and how diet contributes to disease risk.

Chapter 4
You are what you eat

Introduction

In the Martyrs' Museum in the village of Tolpuddle in Dorset, England, there is a chart that summarizes the 'wages of despair'. In the early 19th century, the average Dorset farm labourer earned between 9 and 10 shillings a week. Their families lived on a diet of bread, tea, and potatoes, with small amounts of cheese, sugar, and salt. The weekly household bill for this very basic diet, plus rent, fuel, and other necessities, came to 13 shillings and 9 pence. Farm labourers' families in Dorset were literally starving and in 1834 six of them signed an oath to protest about their wages. They were arrested and deported to Australia; subsequently they became known as the Tolpuddle Martyrs, and after three years they returned, pardoned and to a hero's welcome, as a result of a public petition signed by 800,000 people.

By the end of the 19th century, Britain was the richest country in the world, with the largest empire in history, but the nutrition of its poorest citizens was still dreadful. The government took notice when it needed to recruit soldiers for the second Boer War, against the Dutch settlers in South Africa, in 1899. Up to 40 per cent of recruits were unfit for service and 80 per cent were deemed unfit to fight. The recruits were also short by today's standards: the average height of eighteen-year-olds was about 64 inches

(162.6 cms) and the minimum height for recruits was lowered from 160 cm to 152.4 cm. The government of the day was shocked into action, and in 1904 an Interdepartmental Committee on Physical Deterioration reported. One of their recommendations was the introduction of free school meals to improve the nutrition of the poorest pupils.

A combination of government initiatives and increasing prosperity meant that the average height of the British male increased by about 4 cm during the 20th century and average life expectancy increased by 25 per cent. This is equivalent to an extra six hours per day! Improved nutrition of the population, and improved scientific understanding of nutrition, played a substantial part in these staggering changes.

Globally, malnutrition is still a major health problem. Today there are about 800 million people in the world who do not get enough to eat, and perhaps a further billion who are malnourished because they do not get the right balance of nutrients. But in developed countries the public health challenge now arises from people eating too much rather than too little, with the consequent rise in obesity, the modern counterpart of malnutrition. In this chapter I will explore both kinds of malnutrition.

The origins of the science of nutrition

For about 1,500 years, our understanding of human nutrition came from the writings of the Roman physician Claudius Galenus, or Galen, who was born in AD 130. Galen believed, as had Aristotle and Hippocrates four and five centuries before him, that everything was made up of four elements: earth, air, fire, and water. Galen applied this principle to the make-up of the human body, or bodily humours, and to food. He thought that bodily humours occurred in four fundamental varieties: hot and moist, cold and moist, hot and dry, or cold and dry. These were called, respectively, blood, phlegm, green/yellow bile, and black bile.

Foodstuffs could be characterized by the same four elements, so fruit, for instance, was considered to be cold and moist while beef was hot and dry. Galen was not a fan of fruit: he claimed that his father lived to the age of a hundred by avoiding fruit altogether, a far cry from the modern view that we should eat 'five portions of fruit and vegetables a day'. Remarkably, given that it was completely without any scientific basis, the Galen school of nutrition recommended that a satisfactory meal should include a balance of the four elements, recognizably similar to modern advice on eating a balanced diet. Although Galen's ideas were modified during the late Middle Ages, it was not until the beginning of the Age of Enlightenment in the 17th century that the scientific study of nutrition began to emerge, albeit slowly and with many false steps.

During the 18th and 19th centuries it was established by Antoine Lavoisier, Claude Louis Berthelot, and others that all living things are made of four principal chemical elements—carbon, hydrogen, nitrogen, and oxygen—and that foodstuffs contain these elements in three classes of chemicals: proteins, the building blocks of muscle, fats, and carbohydrates. It was known that living things also contained very small amounts of other elements, such as sulphur and phosphorus, but the roles and significance of these micronutrients were not well understood. Another advance in the science of nutrition came from understanding how the body metabolizes food. For instance, the 19th-century French scientist Claude Bernard showed that the body breaks down complex carbohydrates such as starch into glucose, one of the simplest sugar molecules, which is then utilized by the body to create energy.

Nevertheless, the pioneers of the chemistry of human nutrition often held serious misconceptions, including Berthelot's belief that humans could absorb nitrogen from the atmosphere. In the 19th century the German chemist Justus von Liebig made a fortune out of 'extractum carnis', a supposedly nutritious meat extract rather like a modern stock cube, of virtually no nutritional value, and a 'perfect infant food' that caused malnutrition because it contained no vitamins.

Vitamins

By the late 19th century it had been established that proteins, carbohydrates, and fats are the main nutrients, along with small amounts of minerals such as phosphorus, sulphur, potassium, and sodium. But vitamins, essential for growth and healthy survival of animals including humans, were still unknown. The history of the discovery of vitamins shows how new scientific evidence can be ignored when it contradicts the established dogma of the scientific and medical establishment. It also shows that in the end, evidence and experimentation are able to overturn received wisdom and prejudice. Over a period of about 150 years from the middle of the 18th century, diseases, now known to be caused by vitamin deficiency and easily preventable or treatable by small changes in diet, were persistently ascribed to other causes, even in the face of convincing evidence for a nutritional basis.

Perhaps the most remarkable story of resistance to evidence is the history of scurvy. Scurvy sufferers have bloody patches under their skin, lethargy, constipation, joint pains, softened muscles, rotting gums, and a stench of putrefaction. Scurvy was a particular problem among sailors at sea, and because of the importance of healthy sailors in naval warfare it was the subject of much investigation.

The 16th-century navigators Pedro Cabral and Sir Richard Hawkins both discovered that they could successfully treat scurvy at sea by giving their sailors citrus fruits, but the mainstream view was that scurvy was caused by factors such as foul air below decks or too much salt from salted meat (a view held by Admiral Nelson 300 years later, right to the end of his life). Even after the Scottish naval surgeon James Lind had carried out a proper, controlled experiment in the 1740s showing that citrus fruit cured scurvy, the

English navy favoured other treatments such as sulphuric acid or malted barley. It is often, mistakenly, said that Captain James Cook discovered that fresh fruit prevents scurvy, but in fact he attributed his success in preventing scurvy to giving his crew a daily dose of malted barley. It was a lucky coincidence that he often landed on his very long voyages to take on board fresh fruit and vegetables. This was the real reason that his sailors did not get scurvy.

Eventually, during the Napoleonic Wars at the end of the 18th century, the evidence for citrus fruit was so strong that it was accepted by the English navy, and by the Battle of Trafalgar in 1805, scurvy had been largely eliminated by dosing sailors with a daily ration of 'lime juice', in fact more often lemon juice, mixed with rum, giving rise to the nickname 'limey' for English sailors. Remarkable though it may seem, this was not the end of the dispute. At the end of the 19th century, the Royal Society set up a committee of enquiry chaired by Lord Lister, the discoverer of antiseptics, that concluded: 'neither lime juice nor fresh vegetables either prevent scurvy or treat it . . . scurvy is a disease produced through eating tainted food.' Why did the Royal Society reach such a bizarrely incorrect conclusion more than ninety years after the navy had agreed that scurvy is treatable by giving sailors citrus juice? The answer is that in 1875, a British Arctic expedition had suffered rampant scurvy in spite of taking ample quantities of lime juice. This seemed to refute the citrus theory. It was only in 1918 that the mystery was finally solved when it was shown that lime juice rapidly loses its anti-scurvy effect when preserved with alcohol, as it was on the expedition. Other similar expeditions had used lemon juice and not suffered from scurvy, and lemon juice retains its activity much longer. We now know that the active ingredient is vitamin C, or ascorbic acid, which breaks down over time.

In an almost uncanny parallel, the discoveries of the ability of three other vitamins to treat or prevent diseases also met with great resistance from the scientific and medical establishment. The diseases are beriberi, rickets, and pellagra. In each case, experimental evidence showing that diet could prevent or treat the disease was ignored in favour of other hypotheses.

The word vitamin was first used in 1912. It was originally 'vitamine' because the man who coined the word, Casimir Funk, thought, incorrectly as it turned out, that all vitamins belonged to a family of chemical compounds called amines, and vitamine was a contraction of 'vital amine'. By this time, scientists had accepted that, in addition to carbohydrate, protein, fat, and minerals, the human body needed other essential nutrients in order to function, namely vitamins. By 1941, thirteen vitamins essential for human health had been discovered.

They play a range of vital roles. For instance, vitamin A is required in the eyes, because the chemical compound in the retina that turns photons of light into electrical signals to the brain is derived from vitamin A. Most vitamins help to power biochemical reactions in the cells of the body: for instance B_1 and B_2 are involved in the chemical pathways that generate energy, and vitamin C is important in the manufacture of collagen, the most abundant protein in the body. It also acts as an 'antioxidant', meaning that it has the potential to protect cells against damage to their DNA (see Chapter 1 and below). The human body cannot manufacture twelve of the thirteen vitamins, so they have to be consumed in food. The exception is vitamin D. While many people get much of the vitamin D they need from food, the body can manufacture it as a result of exposure to sunlight. The ultraviolet light (UVB) in sunlight helps the body to convert cholesterol into vitamin D. Different species of mammals are able to manufacture different vitamins. For instance, amongst the primates, tarsiers, monkeys, and apes, including man, lost the ability to synthesize vitamin C about sixty million years ago. Other primates, the lemurs and lorises, can synthesize it, as can most other mammals. Our remote primate ancestors ate plenty of vitamin C in their diet, and natural selection favoured individuals that did not waste metabolic machinery and energy making a superfluous compound.

Nutrition and health

Scientific understanding of human nutrition is incomparably better now than it was in the 19th or early 20th centuries, and it is now known that in addition to twelve vitamins, there are about seventeen minerals, ten amino acids, the building blocks of proteins, and two fatty acids that we have to consume in our food because we cannot manufacture them ourselves. The science of nutrition forms the basis of nutritional guidelines or dietary reference values (DRVs) on what constitutes a 'healthy diet', for vitamins, minerals, and other essential micronutrients that the human body cannot manufacture, as well as the macronutrients: carbohydrate, fat, protein, and total energy. The recommended intake varies amongst different vitamins and minerals according to their role in the body. For instance, a 70 kg body contains 1 kg of calcium and 3 mg of cobalt, so the amounts of these two minerals necessary for building and maintaining the body are very different.

Serious deficiencies of essential nutrients are relatively uncommon in the developed world today, although there are some exceptions. In the UK, for example, some degree of iron deficiency is seen in 27 per cent of menstruating teenage girls, associated with low consumption of red meat or other sources such as pulses and leafy green vegetables, while at the same time losing iron in menstruation. In the USA, iron deficiency affects between 9 per cent and 16 per cent of teenage girls. Vitamin D deficiency is increasing in developed countries, in part because of changes in lifestyle. As described earlier, an important source of vitamin D is its manufacture from cholesterol through exposure to sunlight. Therefore spending time outdoors reduces the chances of vitamin D deficiency. Changes in lifestyle mean that people spend more time indoors than they used to and furthermore they use sunscreen when they are outside. As a result, rickets, a disease of vitamin D deficiency, although still rare, is increasing among young children in the UK, USA, and other Western countries. Some reports have also suggested links between vitamin D levels in the blood and the risk of certain cancers, as well as multiple sclerosis. The main dietary sources of the vitamin are eggs and oily fish, although some processed foods such as margarine and breakfast cereals are often fortified with vitamin D.

In the developing world, deficiencies in essential nutrients affect very large numbers of people: over 1.5 billion suffer from vitamin A deficiency, causing more than half a million children each year to go blind; zinc, iron, and iodine deficiencies are also widespread. What should be done to alleviate this problem? Some argue that it is largely a matter of poverty, and that once the poorest people have enough money they will be able to buy more nutritious food. Others emphasize that the immediate response should be to distribute pills containing vitamin A or other supplements, or provide fortified foods. Thus about 200 million children a year are given vitamin A pills in developing countries. Still others argue that crop improvement is the only long-term solution: nutrient deficiencies are a result of a monotonous, poor diet, dominated by staple crops that do not contain enough of the essential nutrients for health. For instance, vitamin A deficiency in Uganda has been cut by the introduction of a new variety of sweet potato with increased vitamin A content. As we shall see in Chapter 5, one technology that has great potential for improving the nutritional value of staple crops, including tackling the problem of vitamin A deficiency in the developing world, is genetic engineering.

Diet and chronic disease: epidemiology

Prevention of nutrient-deficiency diseases such as scurvy, pellagra, and rickets is the starting point for a healthy diet, but there is more to it than this. On the one hand, there may be additional benefits of consuming some essential nutrients above the minimum level for disease prevention; on the other hand, consuming too much of certain things, such as salt and saturated fatty acids, is known to have adverse effects on health. Consuming too much energy, which can lead to obesity, is also a significant health risk, and I will return to this later in the chapter.

It is now thought that an inappropriate diet is a significant contributor to the risk of suffering from the chronic diseases that are the major causes of morbidity and premature death in the developed world: cancer, cardiovascular disease, stroke, and type 2 diabetes. Three kinds of evidence have contributed to our understanding of diet and health: experimental studies of chemical reactions in the laboratory (*in vitro* studies), experimental studies of human or animal nutrition, and population-level observational studies which look for correlations between differences in diet and differences in health outcomes.

Each of these approaches has its limitations, and ideally results from all three would be used as an evidence base for a healthy diet. The *in vitro* studies can reveal the precise biochemical mechanisms, experiments can test the significance of these mechanisms in the living body, and observational studies can give an indication of their relative importance in the real world. However, things are not always that simple. Mechanisms shown to work *in vitro* may not work in the same way in living tissue. For example, *in vitro* studies of certain vitamins including vitamins C and E, as well as plant secondary compounds, show that they can prevent oxidative damage, analogous to rusting of an iron nail, to DNA. Therefore in principle they could help to protect the body against cancers that arise from damage to DNA. Indeed it is often asserted that consuming these antioxidants protects against cancer. But in the majority of experimental studies in which people are assigned at random to receive pills containing either antioxidants or placebos, there is no evidence that the antioxidants protect against cancer. These studies cast doubt on the anti-cancer benefits of taking food supplement pills, but they do not rule out the possibility that there is a more subtle effect of eating antioxidants or the foods that are rich in antioxidants. For one thing, the experiments typically last only a short time and it could be argued that diet over a whole lifetime is what matters. Furthermore, the experiments usually only involve providing supplements of a few antioxidants at a time and it is possible that consuming a wider range of these chemicals could help protect against cancer.

Large-scale epidemiological studies of diet and health outcomes, involving many thousands of people, generally fall into two categories: retrospective and prospective. Retrospective studies look back through time to see if there are any associations between the kinds of food people eat and their risk of chronic disease. Prospective studies recruit a large number of volunteers and then follow their fate over many years. The largest and most geographically widespread prospective study is the European Prospective Investigation into Cancer and Nutrition (EPIC) in which more than 500,000 people aged between twenty-five and seventy from ten European countries were followed for over a decade, starting in the 1990s. In the 1990s and early 2000s there were similar, although smaller-scale, prospective studies in North America and other parts of the world. Epidemiological studies have a number of limitations: they may show correlations between diet and health outcomes, but correlation is not necessarily the same as causation; they cannot reveal the underlying cellular and biochemical mechanisms for links between diet and health; and they have to try to eliminate possible 'confounding effects'. For example, if people who eat fewer fruit and vegetables are also more likely to smoke and less likely to take exercise, both risk factors for cancer, the study must separate out the effects of diet from these other effects. Furthermore, eating less of one kind of food, such as vegetables, might be associated with eating more of another, such as chocolate. Disentangling these confounding effects is achieved by one form or another of a statistical technique called multiple regression analysis that enables the investigator to look for an effect of, say, fruit and vegetable consumption, while holding the other factors constant. Teasing out these effects by statistical techniques usually requires a large number of observations, so the most useful epidemiological studies are those that have involved large numbers, preferably tens of thousands, of participants.

A further limitation of the epidemiological approach is that it relies on people recording accurately what they have eaten as well as their other lifestyle behaviours. Detailed studies of energy intake have shown that people are highly inaccurate at reporting what they have eaten. For all of these reasons, experts generally accept findings of correlations between diet and health outcomes only after they have been independently verified in several epidemiological studies. The results from different studies can be statistically combined in a so-called 'meta-analysis' to see if there is a consistent pattern. Ultimately, however, only experimental studies in which different diets are allocated to different participants can give conclusive evidence.

The earliest systematic analyses of diet and health, in the middle of the last century, were retrospective studies, often comparing different countries with notable differences in the incidence of cancer or cardiovascular disease. One especially striking observation was that in Mediterranean countries, heart disease, as well as various cancers, and perhaps diabetes, was relatively less common than in northern Europe. As a result of statistical analysis taking into account other confounding factors such as drinking, smoking, and exercise, this was attributed in large part to diet. The Mediterranean diet included relatively small amounts of saturated animal fat and red meat, but large amounts of complex carbohydrate (starchy foods such as bread, pasta, or rice), fruit and vegetables, fish and unsaturated vegetable oils such as olive oil, and moderate amounts of red wine. An experimental study called the 'Lyon Heart Study' followed people who had suffered a heart attack and were assigned at random to either a Mediterranean diet or their normal diet: four years later the former group had experienced between 28 per cent and 53 per cent lower risk of recurrence of heart attacks.

The conclusions of a large number of studies of the Mediterranean diet have had an important role in influencing dietary advice in many countries. We are encouraged to eat more fish, fruit and vegetables, fibre (in fruit, vegetables, and whole grains), and complex carbohydrates, less saturated fat, red meat, and fewer high-fat dairy products. Other epidemiological studies have also found links between diet and health risk. The most consistent, and therefore reliable, results are as follows. High salt intake is linked to an increased risk of high blood pressure and therefore to the risk of heart disease or stroke, and possibly stomach cancer. Low fibre intake is linked to an increased risk of bowel cancer. High levels of red meat consumption are linked to a higher risk of bowel and stomach cancer, and high levels of saturated fat consumption are linked to risk of cardiovascular disease. This does not mean that people should entirely avoid red meat, salt, saturated fat, and so on. In fact, modern dietary advice is not dissimilar to that of Galen: moderation and balance in everything.

From epidemiological studies it has been estimated that about 25 per cent of the variation in risk of suffering from one form or another of cancer or from cardiovascular disease is related to diet. The EPIC study referred to above, for instance, found that eating one extra portion of fruit or vegetables per day (80 g) reduces risk of cardiovascular disease by 4 per cent.

Health claims for foods

Although the evidence relating diet to risk of chronic disease is by no means complete, experts agree that what we eat has a significant effect on our risk of becoming ill or dying prematurely, and the advice on what constitutes a healthy balanced diet is similar in all developed countries. But there are also plenty of reports in the media, and on the Web, of claims for the health benefits of particular foods. It may be 'super foods' such as blueberries, broccoli, or chocolate that prevent cancer, fish oils that make children more intelligent, vitamin supplements that will help stave off symptoms of the menopause—or more outlandish advice, like the claim that green vegetables will oxygenate your blood.

For the most part these claims are not supported by evidence. Unless a result has been verified in more than one study, properly conducted on a large sample size, it is best to treat it with extreme scepticism. For instance, the claim that blueberries are a super food rests on their content of antioxidants. But as we have seen, the role of antioxidants in preventing cancer in the living body as opposed to the test tube is not clear, and in any case many other fruit and vegetables are equally rich in antioxidants. Even if antioxidants turn out to have a protective effect, it does not necessarily follow that more is better: there could be a threshold dose above which there is no more added benefit. Proponents of organic food have claimed that it is more nutritious, for instance because it contains more antioxidants. However, many studies have failed to reveal consistent differences in nutritional value between organic and conventional food.

The food supplements industry makes a lot of money by selling pills that contain trace nutrients or other substances that have supposed health benefits, even when no such benefit has been conclusively demonstrated. The wording of these claims is carefully crafted, for instance by using phrases such as 'believed to' or 'has been reported to'.

In 2010, the global market for fish oil pills was worth about $2 billion. Cod liver oil, along with oily fish such as mackerel, sardines, salmon, and fresh tuna, are good sources of the long-chain polyunsaturated fatty acids that are important components of the human body, especially the nervous system, including the brain. Saturated fats tend to be solid at room temperature, and most animal fats are saturated, as are some vegetable fats such as palm oil, as well as synthetic 'transfats'. It is thought that most of these fats, when consumed in large quantities, increase the risk of heart disease. Unsaturated fats such as olive oil are liquid at room temperature and some of them are linked to a reduction in the risk of heart disease. The two unsaturated fatty acids that the human body cannot manufacture, both 18 carbon atoms long, are called alpha-linolenic acid (an omega-3 fatty acid) and linoleic acid (an omega-6 fatty acid).

In the body, these two essential fatty acids are converted into longer-chain molecules that are important in the nervous system and the functioning of the heart and circulatory system, as well as in anti-inflammatory responses of the body. The longer-chain molecules are the ones that are particularly abundant in fish oils. Although some plant oils such as sunflower, canola, and pumpkinseed contain the essential fatty acids, their conversion to the longer-chain varieties in the body is not efficient, and therefore fish oil is the best source of the fatty acids the body needs. There is good experimental evidence from trials in which people are given fish oil pills that these can help prevent recurrence of heart disease, and some indications that it can prevent dying from a primary heart attack. Hence people are advised to eat oily fish at least once a week.

But what about the effect of fish oils on the brain? There are claims that giving children fish oil pills improves their academic performance. Sixty per cent of the dry weight of the brain is fat and the most important fats are the long-chain polyunsaturated fats found in fish oils. These fats are critical in brain development, but that does not mean that eating more will make children more intelligent or better at concentrating and there is no consistent evidence to support these claims. It has been suggested that fish oil supplements might help to treat behavioural impairments such as dyslexia, dyspraxia, and ADHD (attention-deficit hyperactivity disorder), but not all studies show a beneficial effect.

Nutrition and development: intergeneration effects and epigenetics

It is not only an individual's diet that influences the risk of succumbing to the chronic diseases of developed nations. The mother's nutrition also has a significant effect. David Barker, building on earlier work, discovered that the lower a baby's birth weight, the higher the risk of heart disease, stroke, and hypertension in later life. He hypothesized, and this is now widely accepted, that if the mother suffers from poor nutrition during pregnancy, this influences her ability to transfer nutrients to the foetus, which in turn affects the offspring's long-term health prospects. Barker also found that if underweight babies rapidly gain weight after the age of two, their risk of later heart disease is increased. Some of these effects are direct: undernourished foetuses have less heart muscle and so less ability to survive damage to the heart. They also have smaller livers, and the liver controls blood cholesterol level, which in turn affects risk of heart disease.

But there is also another mechanism that has been uncovered in recent decades. Genes in the offspring may be turned on or off by environmental influences in the parent: a process called epigenetics. So the infant of a poorly nourished mother might have its genes programmed in such a way as to increase its susceptibility to diseases in later life. The Dutch famine of 1944–5, which resulted from a German food embargo, showed the importance of maternal nutrition. Babies born to mothers who were malnourished in the first three months of pregnancy during the 'hunger winter' were more likely to become obese, and to suffer from chronic diseases, in later life. This could, in part, have been a result of epigenetic effects.

Epigenetics plays a role in an even more remarkable discovery, namely that the nutrition of grandparents affects the life chances of their grandchildren. One study examined the life expectancy and disease susceptibility of people born in the northern Swedish town of Øverkalix in the early years of the 20th century. Their grandparents had lived through one or more of the 19th-century famines. The results suggested that the sons and grandsons of men who had gone from famine to feast between the ages of nine and twelve lived, on average, six years less than the comparison group. There was a similar relationship between grandmothers, daughters, and granddaughters. You are what your ancestors ate.

The precise details of how this transmission through two, possibly more, generations works are not yet known, but it probably involves epigenetic programming of genes that affect disease susceptibility and life expectancy. The fact that the effect is transmitted through fathers as well as mothers shows that it must be an epigenetic effect, because fathers hand on only DNA to their children, whereas mothers pass on both DNA and nutrients in their eggs. A woman's eggs are all formed before she is born, so with females there could be a direct effect as well as an epigenetic effect: poor nutrition of the mother could result in some change in the content of the eggs of her unborn daughter.

In many developed countries today there are large differences in life expectancy and disease susceptibility between the wealthiest and poorest members of society. In the UK, for instance, life expectancy varies by about thirteen years between the richest and poorest people. Much of this difference may be caused by direct effects of poverty on housing, diet, and lifestyle. However, it is also possible that intergenerational epigenetic effects play a role, making the elimination of health inequalities an even greater challenge than it might appear at first sight.

Nutritional wisdom

Surveys of pregnant women reveal that they often experience food cravings. Different individuals report cravings for sweet foods, pickles, citrus fruit, salty foods, and even clay. It is sometimes said that these cravings reflect nutritional needs during pregnancy and that expectant mothers are displaying 'nutritional wisdom' by seeking out foods to meet their dietary needs. There is no convincing evidence for this idea and the mechanisms by which deficiencies of different nutrients might be detected have not been identified. The most convincing demonstration of nutritional wisdom is from studies of rats in which it was found that they are able to detect deficiencies in the ten essential amino acids (see above) in their diet: if just one is missing from their diet they will stop eating that particular food in about half an hour. The sensors that detect the amino acid deficiency are in the region of the brain that processes scent, the olfactory cortex.

Obesity: the new malnutrition

During the late 20th and early 21st centuries a new global trend has emerged: people are getting fatter. The increase in the prevalence of obesity started in the rich countries, but has spread to middle- and low-income countries. Generally, in rich countries obesity is more prevalent among poorer people, while the reverse is true in low-income countries. The World Health Organization estimated in 2011 that about 1.5 billion, or 30 per cent of, adults aged twenty or over in the world were overweight or obese, and of these 0.5 billion, or one in ten adults, were obese. The equivalent estimates for school-aged children are 200 million, a little under 10 per cent of the total, of which 40–50 million are obese. In the European Union, 60 per cent of adults and 20 per cent of children are overweight or obese. The obesity crisis is sometimes described as a global disease pandemic, although some would argue that it is not a disease in the usual sense, even though there is little doubt that it is associated with many forms of illness.

There is significant variation between countries in the prevalence of obesity. At the top of the league, excluding some very small countries, is the USA, with around a third of adults obese and over one-third of children obese or overweight. In many rich countries, including Canada, Australia, New Zealand, and the UK, a quarter to a third of adults are obese. In these same countries a quarter to a third of children are overweight or obese. In Europe, however, Greece tops the children's league with 37 per cent of girls and 45 per cent of boys being obese or overweight. Obese and overweight children are also becoming a problem in middle- and low-income countries such as Bolivia, Chile, and Mexico, as is adult obesity in Egypt, Paraguay, and Venezuela. In most countries there are small gender differences, but in some Middle Eastern and African countries, including Turkey, Saudi Arabia, and South Africa, the prevalence is much higher among women than men. One problem with these comparisons among countries is that the methods of data collection vary, so the numbers must be treated with a degree of caution.

The increase in the prevalence of overweight and obese people in the late 20th and early 21st centuries is staggering. It is perhaps best illustrated by the figures for children, because for adults, the increasing age profile of the population could be a confounding factor, as people tend to put on weight as they get older. Illustrative examples are the USA, UK, Spain, and New Zealand, where the prevalence in boys in 2006 was more than double pre-1990 levels, and Brazil, where the increase was around fourfold. Similar changes were seen in girls at the same time. Various studies have projected trends in obesity into the future, including a UK government study that estimated 40 per cent of adults could be obese by 2025, without effective policies to stem the rise.

What is the definition of obese and overweight? The internationally agreed measure is body mass index (BMI), calculated as weight in kg/height in m^2. The normal range is considered to be 18.5–24.99; overweight is 25–29.99 and obese is >30. There are other measures that are also used such as waist-to-hip ratio and waist circumference. These may be better predictors of some of the health risks associated with obesity than is BMI, because they take into account the way in which excess fat is deposited in the body.

The primary reason for concern about obesity among health professionals and politicians is the associated health risks. Comparisons between overweight or obese and healthy-weight people show that the risk of many non-communicable chronic diseases, including a wide range of cancers, cardiovascular disease, diabetes, infertility, and respiratory disease, is substantially increased as a result of being too fat. The increase in risk for some of these diseases, such as diabetes, is over 40 per cent. Excess body mass is said to be the second largest preventable cause of premature death in the USA, behind smoking, and the World Health Organization estimates that 65 per cent of the world's population live in countries where being overweight or obese are bigger causes of death than is shortage of food. In addition to the individual suffering, the disease burden associated with overweight and obesity will cost health-care systems very large sums of money. According to one estimate, the cost will be around \$50 billion per year in the USA and £2 billion per year in the UK by 2030.

At one level, the explanation for the dramatic rise in obesity is very simple. If people consistently eat more calories than they expend, the extra calories are stored as fat, so the rise in BMI is because people are either eating more or spending less energy, or both. However there is considerable debate about precisely what mixture of energy intake and energy expenditure is responsible, and about what the mechanisms are that lead some people but not others to become obese.

The food industry, perhaps not surprisingly, usually points the finger at changes in lifestyle that make us all less likely to spend energy. Many people have sedentary jobs sitting at computer screens, many drive rather than walk or cycle, take the elevator instead of the stairs, and watch TV, surf the Web, or play computer games instead of getting out and playing sport.

On the other hand, there is near consensus among obesity experts that a major factor in recent years is the dramatic increase in the availability of cheap, energy-dense food that enables, or even encourages, people to eat more calories than they need. A stroll down the main street of almost any town or city in the rich countries is enough to remind us of the abundance of fast food outlets. David Kessler, former head of the US Food and Drug Administration, has gone further, even suggesting that modern fast food is addictive and stimulates the same forebrain neural circuits, involving the neurotransmitter dopamine, that are associated with drug addiction. This is speculative, but it is very likely, as we saw in Chapter 1, that we have evolved to like fat, sugar, and salt. These are the key ingredients that the food industry has artfully distilled into fast foods, along with other attractive features that contribute to flavour such as mouth feel, crunchy sounds, and visual appearance (see Chapter 2). This is why milk shakes, doughnuts, and French fries are so irresistible. The energy density of fast food and snacks also means that we are able to consume a lot of calories before our stomach feels full: a small, 25 g bag of crisps contains about the same number of calories as 500 g of French beans. A further factor may be the protein content of fast food. Fast foods tend to contain relatively little protein and this means that they do not quench hunger effectively: protein has been shown to quench hunger for longer than carbohydrates and fats.

It is surprising how little extra energy intake results in weight gain. The increase in the average weight of an American adult over the past thirty years is 9 kg, which requires consuming only ten extra calories per day, the equivalent of eating 1/20th of a glazed doughnut or drinking 1/20th of a tall latte. Many people eat or drink these as snacks between meals without even thinking of them as food. This in part explains why most of us under-report what we have eaten. People of normal weight under-report by about 20 per cent while those who are obese do so by about 30 per cent.

On the other side of the equation, it is very hard to lose weight by exercising. Twenty minutes on the treadmill at the gym at 12 km/h uses up about the same number of calories as are in a 330 ml bottle of soft drink. Eating less is also slow to produce results: in order to lose 11 kg over three years, you would have to forsake 250 calories a day, equivalent to a gin and tonic with a handful of peanuts, a 50 g bar of chocolate, or a cola and crisps. There are many diet regimes that claim to be the magic bullet for weight loss, but in the end the only recipe that works is to eat fewer calories than you expend, and to sustain this change in habit.

Irrespective of the exact balance between intake and expenditure, obesity is the result of the body's normal regulatory mechanisms being overridden. Our brain is equipped with sophisticated feedback mechanisms to enable us to match our intake with our expenditure, and for most of human history, for most people, these homeostatic controls have worked well. But the modern, 'obesogenic' environment with its almost endless availability of cheap, varied, attractive calories and diminished need to expend energy, extends outside the range of conditions for which our feedback mechanisms work effectively. There is a mismatch between our physiology and our environment. But not everyone is obese, so clearly there is variation in propensity to respond to an obesogenic environment by putting on weight. Some people are lucky, because of genetics, epigenetics, upbringing, or a mixture of these, and can cope. Those who cannot are not getting fat because they want to: far from it, most overweight and obese people would prefer to lose weight. In some countries, the rise in obesity is reported to be slowing. This may be because the most susceptible individuals have become overweight or obese, and the remainder are more resistant to the epidemic, rather than any direct effect of policies.

The problem of obesity and its associated health risks is now universally recognized as a major and urgent public health challenge, but there has been remarkably little effective action. The reasons are simple. First, there is no simple magic bullet, analogous to a vaccine, which would tackle the problem. Second, some of the effective actions involve regulating the food industry, and governments are reluctant to take them on, as they were with the tobacco industry for many decades. Third, interventions that would restrict choice would be unpopular and be seen as the worst excesses of the 'nanny state'. So most governments rely on voluntary agreements with the food industry and programmes of information or other ways of attempting to change peoples' eating and exercise habits. Sadly, the evidence is that these tactics have limited effect.

The evidence concerning which policies could make a difference is far from conclusive. However, restricting promotion and advertising of high-calorie food and drink to children is likely to have a positive benefit. The results of research on labelling processed composite foods using 'traffic lights' with red signalling 'high', amber 'intermediate', and green 'low', for calories, fats, and sugar are inconclusive. One review has suggested that a 'junk food' tax rate of 20 per cent could have a significant impact on patterns of food consumption. Finally, there are indications that reducing portion sizes could help to cut consumption of calories.

On a planet where over 1.5 billion people are either short of food or short of some important nutrients, while 1.5 billion are eating too much, one might conclude that a simultaneous answer to the problems of obesity and hunger would be to transfer food from the places that have too much to places with too little, but this may not be realistic, so in the final chapter I turn to the question of whether, in the future, there will be enough food to go round and bring an end to hunger.

Chapter 5
Feeding the nine billion

Introduction

In the time it takes to read this chapter, say twenty minutes, there will be around 3,600 hungry new mouths to feed. Every second, about six new babies are born and three people die: so the world's population is increasing at about 180 mouths per minute. Demographic projections are notoriously difficult to get right. However the experts agree that by 2050, there will almost certainly be over nine, and perhaps as many as ten, billion people on the planet, compared with 6.8 billion in the early 21st century.

Will there be enough food to go round when there are nine billion of us? As we shall see in this chapter, there is no simple 'yes or no' answer.

The idea that the expanding human population of Planet Earth will eventually run up against environmental limits was first proposed in the late 18th century by Thomas Malthus, in his famous book *An Essay on the Principle of Population*, published in 1798. He argued that the human population would increase at a faster rate than food production. This is how Malthus summarized his ideas:

> The constant effort towards population...increases the number of people before the means of subsistence are increased. The food

therefore which before supported seven millions must now be divided among seven millions and a half or eight millions. The poor consequently must live much worse, and many of them be reduced to severe distress…The power of population is so superior to the power of the earth to produce subsistence for man, that premature death must in some shape or other visit the human race.

Malthus's idea is related to the ecological concept of 'carrying capacity', the maximum population that the environment can sustain indefinitely.

Critics of Malthus observe, however, that while the human population has increased more than sixfold since the late 18th century, food production has increased even more rapidly, so that the amount of food available per person has actually gone up. As we shall see, the magic ingredients that have produced this miraculous result are science and technology, supported by social and economic change. The question is whether or not this will continue to deliver similar results in the future, and if so, at what cost in terms of money and damage to the environment.

In addition to the question of whether the *total* amount of food available will be enough to go round, there is the question of *distribution*. In the first decade of the 21st century there were, paradoxically, two opposing trends. As we saw in Chapter 4, approximately one in six people in the world are eating too much food in relation to their energy needs and becoming obese, while at the same time nearly one in six people in the world are starving and/or eating a diet that is deficient in certain nutrients. So one response to food shortage is to suggest that overfed populations should distribute their food to the underfed.

Improving the distribution of food from places where there is too much to places that do not have enough is an important goal and should be pursued, and it is crucial in responding to acute famines. But there are also reasons for being cautious about it as a long-term solution. First, is it realistic in the longer term? One leading expert put it like this: 'It is a bit like saying there is no problem of poverty in the world, only a problem that some people are very rich and others very poor. This may be true but it is not likely to help to make poor people richer.' There is another, perhaps more fundamental, reason for not accepting redistribution as the sustainable answer, namely that it could promote a culture of dependency on aid, and work against the aim of enabling food-poor countries either to become self-sufficient by boosting their own food production to a point where they can sustain their own populations or to become sufficiently wealthy to buy the food they need from elsewhere.

The potential for more food

All the food we eat ultimately comes from the sun. Plants capture photons of light energy and use them to power a chemical reaction, photosynthesis, that combines water and carbon dioxide into glucose, which in turn provides the energy to support almost every other living thing on the planet. In addition to solar energy, plants need various nutrients. Nitrogen, phosphorus, and potassium are needed in the largest quantities, followed by calcium, magnesium, and sulphur in smaller quantities, and the so-called trace elements needed in tiny amounts. These are zinc, manganese, boron, molybdenum, iron, chlorine, and copper. Without sufficient supplies of nutrients, plant growth is stunted, hence farmers often add them in the form of fertilizers or manure, to which we will return in a moment. The final and crucial ingredient for plant growth is water, the availability of which may increasingly limit future food production in many parts of the world.

Plants capture only a small fraction of the solar energy that falls on them. Crop plants typically have a photosynthetic efficiency of about 1–2 per cent. This is the proportion of photon energy landing on the plant that is turned into growth. About 70 per cent of the theoretically available energy is lost either because it is the wrong wavelength for plants to absorb, or it does not happen to hit a molecule of chlorophyll—the green pigment of plants that captures photons to drive the manufacture of glucose. Much of the remaining 30 per cent of sunlight energy is used up either during the chain of chemical reactions that produce glucose, or for the plant to maintain itself. So one theoretical option for increasing food production might be to develop plants that are more efficient at turning sunlight into potential food for us. Plants have had many hundreds of millions of years of 'product development' by natural selection to improve their efficiency at turning sunlight into growth, so we should not be too confident that we could radically improve photosynthetic efficiency. Nevertheless, there is a major international research effort, funded by the Gates Foundation, to do just this. Under the strapline 'Using the sun to end hunger', the research aims to increase the photosynthetic efficiency of rice, the staple food for half the world's population, by 50 per cent while at the same time reducing the need for water and fertilizers. How might this be achieved? In Chapter 1, I referred to different kinds of photosynthetic pathways: some plants, referred to as 'C_4 plants', have a more efficient way of turning sunlight energy into sugar. Rice is not a C_4 plant, but if the genes controlling the C_4 pathway could be identified and inserted into rice, its efficiency might be increased. However, as I discuss later in the chapter, genetic modification of plants is controversial.

Another way of looking at the potential for food production is to ask how much of the world's current plant growth, usually referred to as 'net primary production', we are using for food. A recent estimate is that the total primary production of the land surface of the planet is 56.8 petagrams of carbon per year (a petagram is one thousand trillion grams, or 1 with fifteen zeros after it). Humans eat about 7 per cent of this and use about another 13 per cent for other purposes such as fuel, buildings, and making paper and fibre. So could we not eat a bit more than 7 per cent? There are at least two reasons why not. First, much of the 93 per cent that humans don't eat is inedible, either because it is poisonous or indigestible; and second, the more primary production that we eat, the less we leave for the rest of nature. There is a trade-off between protecting biodiversity and increasing food production.

Fertilizers

At the turn of the 20th century, agriculture was transformed by the discovery of how to make ammonia out of thin air: the Haber Bosch process that combines nitrogen, 78 per cent of the Earth's atmosphere, with hydrogen. Until the middle of the 19th century there was a dispute about the importance of nitrogen for crop production. The 19th-century German chemist Justus von Liebig incorrectly calculated that plants obtain all the nitrogen they need for growth from rain, and that the key to increasing crop yields was to add minerals including potassium, calcium, and magnesium.

This was shown to be wrong by Sir John Bennett Lawes, an English landowner and entrepreneur, who, in 1843, started the world's first agricultural experiment, which is still running today and is the longest-running experiment of any kind anywhere in the world. Lawes used a piece of land called Broadbalk Field on his family estate at Rothamsted in Hertfordshire, just north of London, to establish plots of land fertilized in different ways. Lawes's experiments, carried out in collaboration with a chemist, Sir Joseph Gilbert, established once and for all that adding nitrogen fertilizer increases crop yields. The plots with added nitrogen, potassium, and phosphorus, or with added farmyard manure, yielded two to three times more wheat per hectare than the unfertilized plot, and this difference has persisted for over 150 years.

Lawes showed that nitrogen fertilizers were crucial for improving crop production, but where was the extra nitrogen to come from? In the middle of the 19th century there were three main sources:

natural mineral deposits, especially nitrates from Chile, ammonia as a by-product of turning coal into coke, and seabird droppings, or guano, from the Pacific coast of South America. All three sources were insufficient in the long run. That is why the Haber Bosch process was the saviour of food production in the early 20th century. The Canadian scientist Vaclav Smil has estimated that only half the world's population at the end of the 20th century could be fed were it not for the Haber Bosch process.

The green revolution

Between 1960 and 2000 the world's population approximately doubled, from three billion to six billion, and yet the amount of food produced per person increased by 25 per cent. This was the result of the 'green revolution'. The productivity of agricultural land in many parts of the world, but especially in Asia and South America, was dramatically increased by a combination of four things: plant breeding to produce better varieties of major crops, irrigation, application of agrochemicals, and mechanization.

In *Gulliver's Travels*, Jonathan Swift wrote of the king of Brobdingnag: 'he gave it for his opinion that whoever could make two ears of corn, or two blades of grass to grow upon a spot of ground where only one grew before would deserve better of mankind and do more essential service to his country than the whole race of politicians put together.' The people who did just this in the middle of the 20th century include Norman Borlaug, who developed high-yielding wheat, and Douglas Bell, who did the same for barley. They, and other plant breeders, created crop varieties that simultaneously put a greater proportion of their photosynthetic energy into the seeds that humans consume, grew on shorter stems and so were less likely to blow over in a storm or collapse under the weight of seeds, and were resistant to diseases called rusts, caused by fungi. This genetic revolution is estimated to have contributed about half of the increase in agricultural productivity of the green revolution. In Pieter Bruegel the Elder's painting, 'The Harvest', the wheat stalks are about the same height as the harvesters, while in a modern wheat field, the mature crop barely grows beyond a farmer's waist. Although people were shorter in 16th-century northern Europe, the main difference is the creation of short-stemmed wheat by selective breeding.

The increases in agricultural productivity that resulted from the green revolution were not spread equally around the world. This can be illustrated by looking at cereal production, including wheat, rice, barley, oats, maize, millet, and sorghum. The biggest changes in productivity were in Asia, where traditional agriculture was transformed by modern technology and total cereal production increased by nearly a factor of three between 1961 and 2001. Most of this increase was accounted for by a two-and-a-half-fold increase in yield per hectare, whilst the area of land under cultivation increased by less than 20 per cent. The consequence is that per capita food production doubled in Asia. At the other extreme, in sub-Saharan Africa farming methods hardly changed. Yields per hectare increased in the same period by a mere 20 per cent, and the area of land cultivated almost doubled in an attempt to meet growing food demand. Per capita food production in Africa did not increase. In Europe and North America, where agriculture was already partially modernized before the green revolution, the increases in productivity per hectare were significant, but less than in Asia. The changes in Latin America were not far behind those in Asia, with per capita production increasing by 1.6 times.

In livestock and dairy agriculture, a combination of breeding, feeding, and disease prevention resulted in a fourfold increase in the number of chickens produced, a doubling in the number of pigs, and more modest increases in grazing animals such as sheep and cows. A 21st-century intensively reared 3 kg chicken grows from egg to supermarket shelf in about forty-five days, compared with over sixty days for a chicken left to forage in the farmyard and fields.

There is no doubt that the green revolution was of huge benefit to mankind in the second half of the 20th century and saved many billions of people in the developing world from starvation. It also resulted in a seemingly inexorable decline in the relative price of food. For example, in the UK, the proportion of household income spent on food decreased from an average of 30 per cent in the middle of the 20th century to less than 10 per cent in the first decade of the 21st century. Many foods that were luxuries in developed countries are now cheap enough to eat every day.

However, the green revolution has also come with a cost.

First, it uses up scarce natural resources such as water. Seventy per cent of the world's available fresh water is exploited by man, and agriculture accounts for about two-thirds of all water use. Overall, the amount of fresh water available per person at the end of the 20th century was only half that in the middle of the century, and about one-third of the world's population was short of water. Eighty-six per cent of the water used is 'invisible' water associated with production of food. It takes about 70 litres of water to grow a single apple and about 15,500 litres of water to produce a kilogram of beef.

Second, intensive agriculture uses large amounts of energy in fertilizer production, pumping water for irrigation, and farm machinery. In natural populations of animals, the number of food calories obtained per calorie expended in gathering food, a measure of efficiency, is typically about ten, with a range from about two to about twenty. Some animals, such as hummingbirds, spend a lot of energy but eat energy-dense food, whilst others, such as sit-and-wait predators, spend little energy but consume less energy-dense food, so the ratio of income to expenditure is similar. Human hunter-gatherer societies, as well as those with primitive agriculture, have similar energy efficiencies: a typical figure of around ten calories eaten for every calorie spent in getting the food to the table. But with modern, intensive food production, taking into account all the fossil fuel calories used, every calorie on the plate requires ten calories of energy input. In other words, while we enjoy a plentiful supply of cheap food, the energy cost of producing each calorie is roughly one hundred times larger than the cost per calorie for our hunter-gatherer ancestors.

Third, intensification of agriculture, by squeezing more out of the land for our consumption, takes away food for the rest of nature, and this is seen in the rapid decline in abundance of wild species of plants and animals in intensively farmed landscapes. In the UK, for instance, formerly common farmland birds such as the skylark declined by about 70 per cent in the period between 1960 and 2000.

In addition to these three costs, the productivity gains of the green revolution may be showing signs of running out of steam. Between 1961 and 1990, rice yields grew at more than 2 per cent per year, but between 1990 and 2007 the growth was only 1 per cent per year and for wheat the comparable figures are 3 per cent and 0.5 per cent. The gains of improved irrigation, fertilizer application, and genetics were large at the beginning, but it became harder to make further gains once productivity had been raised substantially from the starting point. In the first decade or so of the 21st century, food prices began to rise for a complex mixture of reasons, but almost certainly in part because the gains of the green revolution were no longer keeping pace with increasing demand. It is not yet clear whether this trend is inexorable: some would argue that with more investment in research and development, there could be a new surge of increased productivity.

No one doubts that the world's demand for food is going to increase sharply in the decades ahead. Experts estimate that the demand will have increased by 50 per cent in 2030, and possibly doubled by 2050. This is a result of two things: the increasing number of mouths to be fed, and the increasing affluence in the developing world, which means that people eat more and eat kinds of food that are more costly to produce, in terms of water, energy, land, and agrochemicals. As countries become wealthier, their populations tend to undergo a nutritional transition from a diet that is often largely plant based towards an increased consumption of meat, especially beef. Each kilo of beef requires about 8 kilos of plant food, and fifteen times as much water as a kilo of wheat needs. One estimate is that by 2050 the average per capita consumption of meat will rise from its current level of about 38 kg to 52 kg per person per year.

Some experts talk about a 'perfect storm' affecting food production during the first half of the 21st century. Demand is going up because of population growth and the nutritional transition, resources such as suitable land, but especially water, are increasingly in short supply, energy costs are rising, and the gains of the green revolution are beginning to slow down. On top of this is the challenge of climate change.

Climate change and food production

It is now generally accepted that human activity, particularly burning fossil fuels with the resulting release of carbon dioxide, has changed the composition of the Earth's atmosphere. As a result, the average global temperature is increasing. Climate models, based on understanding of the underlying physical processes, predict that the average global temperature will rise by at least 2°C by the end of the 21st century, and possibly by much more if greenhouse gas emissions are not curbed. Man-made climate change is also likely to result in a greater frequency of extreme weather events such as floods and droughts. Predictions of the future climate in different regions of the globe are currently less accurate than are forward projections of global averages, but this may improve in the future.

As we saw in Chapter 1, there have been major fluctuations in the Earth's climate in the past, caused not by human activity but by natural phenomena such as variations in the Earth's orbit around the sun. The 10,000 years of agriculture have, however, been a period of climatic stability, so future changes to the climate could have significant effects on agriculture. The impact of climate change on the future of food production is twofold: food production generates greenhouse gases that cause global warming, and climate change itself will reduce agricultural productivity in some parts of the world.

At the start of the 21st century, agriculture and changes in land use accounted for about 25 per cent of global greenhouse gas emissions, but in 2050, with savings elsewhere, they could account for more than 75 per cent if nothing were done to reduce the agricultural contribution. Changes in land use are an important contributor to agricultural emissions, especially chopping down forests to make way for crops, which releases into the atmosphere large amounts of carbon that have been stored in the trees and soil. In short, whatever methods are deployed to increase food production will simultaneously need to reduce greenhouse gases if we are to curb climate change.

The second impact of climate change is that agricultural production itself will be affected by the changing climate. There are likely to be winners and losers. Warmer summers and higher carbon dioxide levels mean that productivity in high latitudes, for example Canada and northern Europe, may increase. On the other hand, currently fertile areas such as the countries of southern Europe may become too hot and arid for growing food. Africa, with its growing population and relatively low agricultural productivity, is predicted to suffer in many parts from more erratic rainfall and long periods of drought. Agricultural productivity may, therefore, decline by somewhere between 5 per cent and 25 per cent, depending on the crop and the location, by the latter part of the 21st century.

The doubly green revolution

To recap, the global challenge for the first half of the 21st century is to produce more food, with more efficient use of energy, water, and agrochemicals, while at the same time reducing greenhouse gas emissions, coping with a changing climate, and avoiding destroying natural habitats and their biodiversity. This has been described as the need for a doubly green revolution, the 'challenge of sustainable intensification' of farming, or the 'sustainable production challenge'. What will it involve? Is it possible to grow more with less?

Different experts emphasize different solutions to the problem of feeding the nine billion: technological, socio-political, and ecological. Although these are sometimes discussed as though they are competing alternatives, in fact they are not. The following sections include a number of examples relating to Africa. The challenge of sustainable intensification is worldwide, but many sub-Saharan African countries have low current agricultural productivity and expanding, increasingly wealthy populations.

Technology

The advocates of a technological solution point to the 'yield gap', the difference between what is actually produced and what could be produced from a hectare of land with the optimum application of existing technologies such as the best genetic varieties, use of agrochemicals, and irrigation. Even in Asia, where yields have increased dramatically, there is often a yield gap of more than 20 per cent. In Africa, where productivity has barely improved at all over the past half century, the yield gap may be as much as 200 per cent to 300 per cent. In Africa, only 4 per cent of crop land is irrigated, compared with 39 per cent in Asia; fertilizer use is less than 10 per cent of the world average of 100 kg per hectare; and at the start of the 21st century only one-quarter of the seeds sown by African cereal growers were improved varieties, compared to 85 per cent in Asia.

Malawi, a small, poor, land-locked African state, has shown how the yield gap can be reduced by a combination of fertilizer application and improved genetic varieties. Maize is the dominant food crop in Malawi: smallholders have an average farm of less than one hectare, and 85 per cent of this is devoted to maize production. From the 1980s onwards, Malawi's farmers could not produce enough maize to feed themselves because of a combination of low productivity, pressure on land, and bad weather. At the end of the 20th century the government of Malawi started to introduce a fertilizer and seed subsidy programme for smallholders. By the end of the first decade of the 21st century, 65 per cent of Malawi's farmers, including many of the poorest, were benefiting from subsidized fertilizer or seed. Maize production doubled from pre-subsidy levels, and Malawi became a maize exporter rather than relying on imports and aid or having its people suffer starvation. Child mortality had halved, probably in part due to greater availability of food. Although the subsidy cost around 9 per cent of government spending, it saved on costly food imports: every dollar spent on subsidy produced over three times as much food as a dollar spent on imports.

Critics of the fertilizer subsidy programme, in addition to pointing to inefficiencies and corruption, have argued that it creates some of the problems of the green revolution, such as its dependence on energy-intensive inputs and production of greenhouse gases that will exacerbate climate change. Other approaches, such as genetic improvement of African crops such as sorghum, millet, cassava, and banana, do not have these disadvantages and would make a major contribution to improving productivity, as would low-tech irrigation techniques that harvest rainwater, such as planting individual trees in the Negev Desert in shallow pits a few metres across, called *Negarim*, to collect rainwater or dew.

Harvard-based Kenyan political scientist Calestous Juma has argued that Africa could feed itself within a generation if it embraced new technologies, rather than adopting the dated technologies of the green revolution. He suggests that Africa should leapfrog to the next generation of technologies, including biotechnology, information technology, geographic information systems (digital spatial information), and nanotechnology. Already, many rural people in Africa have jumped straight to mobile phones without ever having used the 20th-century technology of landlines, and similar jumps could be made in agriculture. Precision agriculture, using satellite imaging and global positioning systems, can allow farmers to target their fertilizers and pest control chemicals in precisely the amounts and locations where they are needed. The Internet could give farmers instant access to advice, weather forecasts, disease spread, and other crucial information. Nanotechnology, the manipulation of materials on the scale of individual molecules and atoms, is able to provide better pesticides and more effective ways of processing food. But the most controversial technology is biotechnology, to which I will return.

Socio-politics

Some people see the problems of food security and food poverty as being essentially socio-political issues to do with rights, ownership of land, social inequalities, and access to international markets, which are often restricted by trade tariffs or subsidies in rich countries. In many African countries, women play a key role in subsistence farming but do not enjoy equal rights or access to essential materials. For instance, one analysis of the Malawi fertilizer and seed subsidy programme noted that it was not sufficiently accessible for women. Another important sociological issue is land ownership rights. If the poorest subsistence farmers are to be self-sufficient and to be able to benefit from new technologies, they have to enjoy security of ownership of their land. Linked to this is the debate about whether the best way to improve agricultural productivity in Africa is to enable smallholders to produce sufficient for their own needs, in other words to modernize subsistence farming, or whether, as the development economist Paul Collier has argued, the best strategy is to develop large agribusinesses to produce for export large quantities of whatever crops grow best in the country. The experience of Brazil shows that both approaches can lead to increases in productivity. These social and economic issues are a key part of the story, and they need to be addressed, but not as an alternative to technological advances.

But new technologies will not be utilized by farmers in the poorest countries until farmers have the capacity to use them, and this in turn depends largely on aid from richer countries being used to develop research, technology, and training in developing countries. Perhaps the biggest change that needs to take place in securing increased productivity in Africa is to reverse the steep decline in agricultural development aid. Between 1983 and 2006, the proportion of aid to poor countries that was spent on agricultural development fell from 20.4 per cent to 3.7 per cent. The USA's proportion of aid going to agriculture fell from 25 per cent to 1 per cent in the same period. This investment is needed both for research and development and for rural infrastructure.

Ecology

The green revolution was, in one sense, about conquering nature rather than working with the grain of nature. It involved eliminating undesirable pests, fertilizing infertile soils, irrigating drought-stricken land, and using big machines to drain, level, harvest, and transport. It also ignored the 'externalities', that is to say, costs imposed on the environment such as pollution of water, destruction of natural habitats, loss of biodiversity, erosion of soils, and contributions to climate change.

Therefore one important line of argument is that food production in the future must work more effectively with nature, and minimize the externalities. What does this mean in practice? What it cannot mean is a decline in productivity per hectare, because if less is produced for each hectare of farmed land, more natural habitat will have to be turned over to food production. This would be bad for biodiversity and would release carbon dioxide stored in soils into the atmosphere as trees are chopped down or grasslands ploughed up. In fact, productivity per hectare has to increase. There are ways that could achieve both aims at the same time: sustainable intensification. These include 'integrated pest control', using a judicious mixture of natural predators and parasites to control pests with small quantities of pesticide where necessary, smart irrigation systems such as the *Negarim* described earlier, and planting patchworks of different crops to both discourage pests that thrive in monocultures and encourage biodiversity. Precision agriculture and biotechnology will also help to reduce inputs of agrochemicals and therefore damage to the environment.

Organic farming aims to produce food in a more 'natural' way, and therefore appears consistent with an ecological approach. However, organic farms have a lower yield per hectare (typically about 60 per cent of that of intensive farming) and so do not offer a plausible way of feeding the nine billion. Importantly, becauses organic farming produces less per hectare, it involves using more land, and ploughing up uncultivated land for farming releases a lot of carbon ino the atmosphere, contributing to climate change. Nor is organic farming necessarily better for biodiversity. Organic farms appear to have higher biodiversity per unit area, although it is not certain that this is a result of lower intensity or something specific to organic techniques. But if looked at in a different way, organic farming may be worse for biodiversity. By using more land, organic farming aims to share food production with biodiversity. An alternative approach of farming some parts of the land intensively and sparing the rest for nature may be better overall. For example, one analysis of butterfly populations on organic and conventional farms in the UK showed that, while organic farms may have greater density of butterflies *per hectare*, when the comparison is based on abundance of butterflies *per unit of food produced*, it is often better to farm intensively and retain some land as nature reserves for wildlife.

Biotechnology

Genetically modified (GM) foods often polarize opinion. For some, GM is both a modern, more sophisticated extension of the process of genetic modification of crops and livestock that started 10,000 years ago and will be a key contributor to sustainable intensification. For others, GM is anathema: dangerous to humans and nature, unnatural, and unnecessary.

Genetic modification of crops involves inserting genes from another organism into the crop plant, with the aim of creating a new variety of the crop plant with a desirable feature such as resistance to insect attack, tolerance to drought, or improved nutritional quality. There are two main methods, both developed in the 1980s, for inserting the 'foreign genes'. For cereal crops such as rice, wheat, and maize, the genes are inserted by coating them on tiny particles of gold or tungsten and literally blasting them into the host DNA with a special airgun. For plants such as potatoes, tomatoes, and sugar beet, the genes are inserted using a bacterium that acts as a Trojan horse. The bacterium, called *Agrobacterium tumefaciens*, is a common parasite of plants and it inserts some of its genes into the host plant in the form of a special fragment of DNA called a plasmid. This plasmid causes the plant to create a tumour or gall, inside which the bacterium thrives. If the plasmid is replaced with DNA that results in a desirable trait for a crop plant, the *Agrobacterium* can be 'tricked' into inserting this new DNA into its host. Both these methods are a bit hit and miss, so only a tiny proportion of attempts to insert new genes into the host produce the desired outcome.

Genetically modified food crops, and some non-food crops such as cotton and tobacco, are grown worldwide on a large and rapidly increasing scale. The use of GM crops for food and cotton increased from 1.7 million hectares in 1996 to over 17 million hectares in 2012. In 2012, GM crops were grown by over 17 million farmers in 28 different countries, and more than 90% of these farmers were poor smallholders in developing countries. One of the key food crops is soya, grown as a source of protein for animals. In 2012, in Argentina 100% of the soya was GM, in Brazil 88% and in the USA 93%. The European Union imports a great deal of soya for animal feed and about 85% of it contains GM or GM derived material. In addition to soya and cotton, the major GM crops grown world wide are maize and oilseed rape (canola). These GM crops have one or both of two traits: resistance to insect attack, which saves the farmers from spraying with pesticides, or tolerance of the powerful herbicide glyphosate, which enables farmers to kill the weeds that could compete with their crops and reduce yields. Some other GM crops, such as papaya, peppers, and squash, are resistant to virus attack. There are also many new GM crops in the pipeline, including foods with additional nutritional benefits such as 'golden rice'. Golden rice contains the precursor of vitamin A, and could help to prevent blindness caused by vitamin A deficiency in about 500,000 children each year in the developing world (see Chapter 4). A bit further down the line are crops that will be resistant to drought or salinity, enabling them to be grown in water-stressed areas, and cereals that are able to fix nitrogen from the air to reduce or obviate the need for fertilizer.

With these benefits, why isn't everyone enthusiastically welcoming GM foods? There is no simple answer. Different groups of objectors, including several environmental non-govermental organizations (NGOs), have different reasons, but in summary they boil down to four main worries: GM may be risky for the environment; it may be risky for human health (this is discussed in Chapter 4); it places too much power in the hands of big corporations that develop and own the intellectual property; it is tantamount to 'playing God' and therefore considered morally wrong.

Some critics also think that the benefits of GM have been exaggerated by biotech companies. The first generation of GM crops was designed mainly to benefit farmers and there has been considerable disagreement about whether the GM crops actually have higher yields or reduce use of agrochemicals. One survey of 168 peer-reviewed comparisons of GM and conventional crops concluded that farmers using GM in developing countries increased their yields by 29 per cent, while those in developed countries increased their yields by an average of 6 per cent. There was wide variation in the results, but in 124 of the 168 comparisons, GM outperformed conventional crops, and in only 13 did they do worse. The review also concluded that insecticide use was 14 per cent to 76 per cent lower on the insect-resistant GM crops and that tillage, which causes soil erosion, was 25 per cent to 58 per cent lower with herbicide-tolerant crops. It concluded that there was an economic benefit to nearly three-quarters of farmers worldwide.

One environmental risk is the possibility that genes for herbicide tolerance or insect resistance could transfer to weed species, creating super weeds. A second concern is that GM may be just one more way of squeezing more out of cropland and leaving less for the rest of nature. A further worry is that insects will evolve resistance to the built-in insecticide of genetically modified crops and weeds will evolve resistance to the herbicide used on herbicide-tolerant crops, as has been reported in some parts of the USA. This is a general worry about agrochemicals and not specific to GM. Some pests, for instance the pink bollworm in India, have become resistant to the built-in insecticide and in other places the fact that farmers with GM crops spray less with broad-spectrum insecticides has led to outbreaks of secondary pests. There are, however, ways of countering the evolution of resistance, such as leaving 'refuges' of non-insect-resistant crop, or developing crops with tolerance of more than one herbicide. Weighed against these concerns are the potential environmental benefits arising from reduced use of insecticides and reduced damage to the soil.

An estimated 75 per cent of all processed foods in the USA contain GM ingredients, mainly from maize, soya, or canola, but in

Europe, consumers have turned against GM foods and no food on sale should contain GM ingredients unless declared on the label. When the first GM food, a tomato paste made from GM tomatoes that were modified to prevent them going soft when ripe, was introduced in 1997, it sold well. But within a year or two, largely as a result of campaigns by pressure groups and some sectors of the media, supermarkets had started declaring that all their products were 'GM free'. Soon after, legislation was introduced by the European Commission, requiring not just GM ingredients but also ingredients *derived* from GM plants, such as soya lecithin or corn syrup, to be labelled. The European consumers' rejection of GM leaves Europe as an outlier, and it has also had knock-on effects in Africa. In 2002, when Zambia suffered a severe drought and its people were starving, the president of the country refused to accept GM maize as food aid, saying, 'Simply because my people are hungry, that is no justification to give them food that is intrinsically dangerous to their health.'

What should we conclude about GM food? The challenge of feeding nine billion by 2050 will need all the technological, sociological, and ecological knowledge that we can generate. GM is not the magic bullet, but it is already having a role to play and its importance may well increase as new genetic modifications come on stream. At the same time, we need to carefully assess the risks as well as the benefits case by case (see Chapter 4 for a summary of the existing regulatory hurdles).

Fish

The oceans cover 71 per cent of the Earth's surface: could marine fisheries make a significant contribution to feeding the world in the future? Fish provide only about 7 per cent of the world's protein and already many fish stocks are over-exploited: according to an estimate by the Canadian fisheries biologist Daniel Pauly, about one-third of stocks are heavily over-exploited and in steep decline, and another third are close to this. Therefore it is unlikely that simply catching more fish will be sustainable. But how about fish farming? Aquaculture contributes nearly half the world's seafood and fish production, primarily in Asia, and there is potential for this to increase significantly. But, as with intensive agriculture, there are also significant problems. These include the environmental damage from pollution by fish sewage or pesticides and habitat destruction to make way for fish farms. Many farmed species, such as salmon and sea bass, are predators and are fed on fishmeal. This means that it takes several kilos of small fish in order to produce one kilo of farmed fish. In short, aquaculture must overcome similar challenges of sustainability as those facing agriculture.

Waste

We waste a lot of food, and reducing waste could make an important contribution to feeding the nine billion. Between 30 per cent and 40 per cent of food is wasted in both developing and developed countries, but for very different reasons. In the developing world the losses are largely on the farm or during transport and storage. Grains such as rice are often devoured by pests or they rot, and fresh produce is lost because neither farmers nor retailers have cold storage. In the developed world, well over half the waste occurs in the home, in restaurants, and in shops, for a combination of reasons. Supermarkets persuade shoppers to buy more than they need by offering what are known as 'bogoffs'— 'buy one, get one free'—or discounts on bulk purchases; fast food outlets offer supersized portions that people cannot finish; consumers are sometimes fussy about the appearance of produce, so cosmetically inferior goods may be thrown away; and 'use by' dates, though useful for consumer protection, are conservative and may mean that perfectly edible food is thrown away. Technology and infrastructure have a role to play in reducing waste, especially in the developing world, but also in developed countries, where new packaging using nanotechnology could increase shelf life. Education, and perhaps legislation, to discourage rich consumers from buying too much might help. Also, if food becomes more expensive over the decades ahead, as seems likely, people in rich countries may, of necessity, become more careful about food waste.

Biofuels and competition for land and water

Before the discovery of fossil fuels, people burned wood, peat, straw, and other forms of biomass, and in the early 21st century an estimated 3 per cent of the world's terrestrial net primary production (plant growth) was used as traditional biomass fuel. But recently a new trend has emerged: turning agricultural land over to the production of liquid biofuels, such as ethanol from sugar cane and maize, or vegetable oils from grain crops. By 2011, 20 per cent of the world's sugar cane and 9 per cent of the world's vegetable oil and coarse grains such as wheat, barley, and sorghum were turned into liquid biofuels.

The move to biofuels is partly driven by concerns about supply of fossil fuels and partly by a largely mistaken belief that they produce smaller quantities of greenhouse gases and will help to mitigate climate change. At first sight it may appear that using crops to make fuel saves carbon emissions, because the carbon released by burning the fuel is recaptured by the next generation of plants. But the story is more complicated than this. First, any calculation of carbon savings should only include any *extra* carbon sucked out of the air by growing biofuel crops: if something else was already growing on the land, there may be no additional carbon saving. Second, growing and processing plants to turn them into fuel often use a lot of energy. Estimates of the total life-cycle carbon savings of various crops for biofuels vary widely, but the overall average is that some crops, such as ethanol from sugar cane, save more than 50 per cent compared with fossil fuels while others, such as ethanol from corn, save very little, if anything. A third point, already mentioned earlier in the chapter, is that when non-agricultural land is converted to cropland, a large amount of carbon stored in the soil as organic matter is released into the atmosphere because when it is exposed to the air by tilling the soil, it is converted into carbon dioxide. Depending on the size of this store, this effect may swamp any carbon savings for years or even decades. Overall, soils contain more carbon than the total amount in the atmosphere and vegetation of the world.

The other concern about biofuels is that they compete with food production for land and water. This is also true of crops such as coppiced willow or the grass *Miscanthus* that are grown as biomass for burning. Part of the rising price of foods in the early 21st century is attributed to this competition. How serious is competition for land? Data on global land use are not always reliable, because classification of land use is complex and often what might appear to be 'unused' land in developing countries is used for low-intensity grazing. However, a rough estimate is that about 12 per cent of the 13,000 million hectares of land on the planet is cultivated, and if grazing land is included, agricultural use accounts for about 38 per cent of land. Much of the rest is unsuitable (mountains, deserts, and so on), protected, or built environment. So-called 'marginal or idle land' accounts for only about 3 per cent of the land area. But even this land may be used by indigenous people for grazing, and there is generally a good reason why it is not used to grow crops: it tends to have poor soils, or is arid. In short, if crops are grown for biofuels, growers will want to use prime agricultural land because this is where productivity is highest, so it is almost inevitable that producing biofuels will compete with food production. This, combined with the fact that the carbon savings are often illusory, is why some experts have argued against expansion of growing crops for biofuels.

The importance of diet

There is an ecological rule called the '10 per cent rule'. In nature, about 10 per cent of the energy in one trophic level is turned into biomass in the next. In other words, 10 per cent of the vegetation eaten by a herbivore becomes herbivore body, about 10 per cent of a herbivore eaten by a carnivore becomes carnivore body, and so on. It might therefore seem an obvious recommendation that people should stop eating meat and switch to a plant-based diet: many more people could be fed off the same area of land if they were vegetarian. If you add to this the fact that agriculture's biggest direct contribution to global warming is the methane that comes out of the rear ends of ruminant livestock such as sheep and cattle, there could be a double gain from eating less, or no, meat: less impact on the environment and less land required to feed the growing human population. Finally, as we saw in Chapter 4, eating too much red meat and animal fat may be bad for our health.

However, the case against meat is not that simple. The conversion efficiency varies greatly depending on the kind of meat and how it is reared: the most efficient is intensively reared chicken. One kilogram of cereal produces 1 kilogram of chicken, for pigs the figure is 4:1 and for cattle, as mentioned earlier, about 8:1. The amount of land needed for a vegetarian diet also varies greatly. One analysis by the Worldwide Fund for Nature concluded that a switch from beef and milk to highly refined vegetarian products such as tofu and quorn could actually *increase* the amount of arable land needed. The carbon footprint of food or other products is usually measured as the 'CO_2 equivalent', which is the global warming potential of various greenhouse gases created during production of a particular kind of food converted into standard units. The footprint of both meat and plant foods covers a very wide span. The UK Climate Change Committee has estimated that sheep meat 'costs' nearly 15 kg of CO_2 equivalent per kg, beef a little less, pigmeat about 4 kg, poultry 3 kg, eggs 2 kg, and milk 1 kg. While vegetables generally have a lower carbon footprint, for instance potatoes are about 0.2 kg, tomatoes grown under glass in the UK fall between chicken and pork. Comparisons based on a kilogram of food, or on calories, are incomplete because they ignore the nutritional composition of the various foods, including protein and trace nutrients. The UK Climate Change Committee estimates that maintaining a similar protein level in the diet but eating chicken instead of red meat could save nearly one-fifth of the greenhouse gas emissions from farming in the UK.

Intuitively one might think that less-intensive farming would have a smaller carbon footprint, but this is not the case for livestock production. In general, the more rapidly the animal grows to the stage at which it is slaughtered for consumption, the less chance it has to generate greenhouse gases on the way. It is also worth remembering that agricultural production up to the farm gate typically contributes less than half the carbon footprint of food. The rest arises from transport, storage, processing, and so on.

If people stopped eating meat, an important consideration for global warming would be what happens to the redundant land. Grazing land is often unsuitable for growing cereals or other crops, and in any case, if it were ploughed up, a large amount of carbon would be released into the atmosphere, adding to global warming.

In spite of these reservations and complications, there is little doubt that by decreasing meat consumption, particularly red meat, the chances of feeding the world in 2050 would increase, and the impact of doing so on the planet would decrease. The problem is that as countries get richer, people tend to switch from a vegetarian diet to a meat-based diet, so achieving a shift in the opposite direction is going to be a real challenge.

Do I worry about whether or not my grandchildren and their children will have enough to eat without further damage to the environment? It will not be easy, but it can be done. The technological answers for sustainable intensification are there, or have the potential to be developed. The greater challenge is the political will to use technology in a wise and judicious way and for every individual to take a responsible attitude to reducing food waste and decreasing meat consumption. If we want to, we can do it.

References

Chapter 1: The gourmet ape

Paleo Diet

Chatham, J 2012. *Paleo for beginners*. Rockridge University Press, Berkeley California

Food history

Harris M and Ross E B (eds) 1987 *Food and Evolution: Toward a theory of human food habits*. Temple University Press, Philadelphia

Evolution of hominins (page 2)

<http://www.nhm.ac.uk/nature-online/life/human-origins/early-human-family/denisovans/index.html>

Hopkin M Ethiopia is the top choice for cradle of Homo sapiens. Nature News. *Nature* | doi:10.1038/news050214-10

Vigilant L et al 1991 African populations and the evolution of human mitochondrial DNA. *Science* 253, 1503–7

Hominin diets studied by isotopes, tooth wear, and tool use (pages 4–6)

Lee-Thorp J 2011 The demise of 'Nutcracker Man'. *Proc Natl Acad Sci USA* 108, 9319–20

Lee-Thorp J 2010 Stable isotopes in fossil hominin tooth enamel suggest a fundamental dietary shift in the Pliocene. *Phil Trans R Soc B* 365, 3389–96

McPherron S, Alemseged Z, Marean C W, Wynn J G, Reed D, Geraads D, Bobe R, Hamdallah A B 2010 Evidence for stone-tool-assisted consumption of animal tissues before 3.39 million years ago at Dikka Ethiopia. *Nature* 466, 857–60

Ungar P S, Scott R S, Grine F E and Teaford M F 2010 Molar microwear textures and the diets of *Australopithecus anamensis* and *Australopithecus afarensis*. *Phil Trans R Soc B* 365, 3345–54

Cooking (pages 7–11)

Pollen M 2013 *Cooked: a natural history of transformation*. Penguin: Allen and Lane, London

Wrangham R 2009 *Catching Fire: How cooking made us human*. Profile Books, London

Brain C K and Sillen A 1988 Evidence from the Swartkrans cave for the earliest use of fire. *Nature* 336, 464–6

Alperson-Afil N, Sharon G, Kislev M, Melamed Y, Zohar I, Ashkenazi S, Rabinovich R, Biton R, Werker E, Hartman G, Feibel C, Goren-Inbar N 2009 Spatial organization of hominin activities at Gesher Benet Ya'aqov, Israel. *Science* 326, 1677–80

Agriculture, domestication, impacts on health (pages 11–15)

Diamond J 1997 *Guns, Germs and Steel*. Jonathan Cape, London

Starling A P and Stock, J T 2007 Dental indicators of health and stress in early Egyptian and Nubian agriculturalists: a difficult transition and gradual recovery. *Amer J Physical Anthrop* 134, 520–8

Larsen C S 2006 The agricultural revolution as environmental catastrophe: implications for health and lifestyle in the Holocene. *Quaternary International* 150, 12–20

The origin of fermentation bacteria (pages 16)

Bolotin A, Quninquis B, Renault P, Sorokin A, Dusko Ehrlich A, Kulakauskas S, Palidus A, Goltsman E, Mazur M, Pusch G D, Fonstein M, Overbeek R, Kyprides N, Purnelle B, Prozzi D, Ngui K, Masuy D, Hnacy F, Burteau S, Boutry M, Delcour J, Goffeau A, Hols P 2004 Complete sequence and comparative genome analysis of the dairy bacterium *Streptococcus thermophiles*. *Nature Biotech* 22, 1554–8

Van de Guchte M, Penaud S, Grimaldi C, Barbe V, Bryson K, Nicolas P, Robert C, Ozias S, Mangenot S, Couloux A, Loux V, Dervyn R, Bossy R, Bolotin A, Batto J M, Walunas T, Gibrat J F, Bessieres P, Weiisenbach J, Ehrlich S C, Maguin E 2006 The complete genome sequence of Lactobacillus bulgaricus reveals extensive and ongoing reductive evolution. *Proc Natl Acad Sci USA* 103, 9274–9

Canning (page 17)

<http://tinplategroup.com/pooled/articles/BF_DOCART_197927>

Chocolate (pages 19–20)

McGee H 2004 *McGee on Food and Cooking*. Hodder and Stoughton, London

Chapter 2: I like it!

Umami and glutamate (page 23)

Sano C History of glutamate production. *Am J Clin Nutr* 2009: 90 (suppl), 728S–32S

Renton A 2005 If MSG is so bad for you, why doesn't everyone in Asia have a headache? <http://www.guardian.co.uk/lifeandstyle/2005/jul/10/foodanddrink.features3>

Different sensory modalities affect flavour (pages 24–25)

Nature Outlook 2012 *Taste* 486, S1–S43

Spence C 2010 The multisensory perception of flavor. *The Psychologist* 23, 720–3

Morrot G, Brochet F, Dubourdieu D, 2001 The colour of odours. *Brain and Language* 79, 309–20.

Anon 2012 Smells like Beethoven. *Economist* 4 February 2012, 74.

Shepherd G M 2012 *Neurogastronomy: How the brain creates flavor and why it matters*. Columbia University Press, New York

Harrar V and Spence C 2013 The taste of cutlery: how the taste of food is affected by the weight, size and shape of the cutlery used to eat it. Flavour 2, 21

Genetic variation in taste (pages 28–30)

Garcia-Bailo B, Toguri C, Eny K, El-Sohemy A 2009 Genetic variation in taste and its influence on food selection. *J Integrative Biol* 13, 69–80

Bartoshuk L 2000 Comparing sensory experiences across individuals: recent psychophysical advances illuminate genetic variation in taste perception. *Chemical Senses* 25, 447–60

Nabhan G P 2004 *Why some like it hot*. Island Press, Washington DC

Early learning and food preferences (pages 30–34)

Rozin P and Schiller D 1980 The nature and acquisition of a preference of chili pepper by humans. *Motivation and Emotion* 4, 77–101

Schaal B, Marlier L and Soussignan R 2000 Human fetuses learn odours from their pregnant mother's diet. *Chem Senses* 25, 729–37

Mennella, J A and Beauchamp G K 2002 Flavor experiences during formula feeding are related to preferences during childhood. *Early Hum Dev* 68, 71–82

Long-delay learning (page 32)

Garcia J, Kimeldorf D J and Koelling R A 1955 Conditioned aversion to saccharin resulting from exposure to gamma radiation. *Science* 122, 157–8

Religious and cultural traditions (page 33)

Harris M 1985 *Good to Eat: Riddles of food and culture*. Simon and Schuster, New York

Vayda A P 1987 Explaining what people eat: a review article. *Human Ecol* 13, 493–510

Harris M 1987 Comment on Vayda's review of *Good to Eat: Riddles of food and culture. Human Ecol* 13, 511–17

Lactase persistence and favism (pages 34–35)

Itan Y, Powell A, Beaumont M A, Burger J, Thomas M G 2009 The origins of lactase persistence in Europe. *PLoS Computational Biology* 5, 1–13

Swallow D M 2003 Genetics of lactase persistence and lactose intolerance. *Ann Rev Genetics* 37, 197–219

Katz S H 1987 Fava bean consumption: a case for the coevolution of genes and culture. Pages 133–59 in *Food and Evolution* ed Harris M and Ross E B Temple University Press, Philadelphia

Spices (pages 36–40)

Jardine L 1996 *Wordly Goods*. MacMillan, London

Sherman P W, Billing J 1999 Darwinian gastronomy: why we use spices. *Bioscience* 49, 453–63

Sherman P W, Hash G A 2001 Why vegetable recipes are not very spicy. *Evol Hum Behav* 22, 147–63

Tomatoes in Italy (pages 40–42)

Gentilcore D 2010 *Pomodoro! A history of the tomato in Italy*. Columbia University Press, New York

Chapter 3: When food goes wrong

Origins of BSE (pages 45–46)

<http://www.archive.defra.gov.uk/foodfarm/farmanimal/diseases/atoz/bse/publications/documents/bseorigin.pdf>

BSE in the USA and Canada (page 47)

<http://www.cdc.gov/ncidod/dvrd/bse/>

Adulteration of food, including milk (pages 48–50)

Wilson B 2008 *Swindled: from poison sweets to counterfeit coffee—the dark history of food cheats.* John Murray, London

Origins of yeast in Bavarian beer (page 49)

Libkind D, Hittinger C T, Valério E, Gonçalves C, Dover J, Johnston M, Gonçalves P, Sampaio J P 2011 Microbe domestication and the identification of the wild genetic stock of lager-brewing yeast. *Proc Natl Acad Sci USA* 108, 14539–44

Toxicology and natural carcinogens in food (pages 51–54)

Ames B N, Profet M and Gold L S 1990 Dietary pesticides (99.99% all natural) *Proc Natl Acad Sci USA* 87, 7777–81

Collman J P 2001 *Naturally Dangerous: Surprising facts about food, health and the environment.* University Science Books, Sausalito CA

Imazilil toxicity (page 52)

Imazilil facts. <http://www.epa.gov/oppsrrd1/REDs/factsheets/2325fact.pdf>

Glycoalkaloids in potatoes (page 53)

<http://www.ucce.ucdavis.edu/files/datastore/234-182.pdf>

Friedman M and McDonald G M 1997 Potato glycoalkaloids: chemistry, analysis, safety and plant physiology. *Critical Reviews in Plant Sciences* 16, 55–132

Acrylamide (page 54)

Tareke E, Rydberg P et al 2002 Analysis of acrylamide, a carcinogen formed in heated foodstuffs. *J Agric Food Chem* 50, 4998–5006

Food additives, and hyperactivity in children (page 55)

<http://www.efsa.eu/en/faqs/faqfoodcolours.htm>

McCann D, Barrett A, Cooper A, Crumpler D, Dalen L, Grimshaw K, Kitchin E, Lok K, Porteous L, Prince E, Sonuga-Barke, E Warner J O, Stevenson J 2007 Food additives and hyperactive behavior in 3-year-old and 8/9-year-old children in the community: a randomized, double-blinded, placebo-controlled trial. *Lancet* 370, 1560–7

Food allergy (pages 56–58)

House of Lords Science and Technology Select Committee session 2006–07 6th report. *Allergy* <http://www.publications.parliament.uk/pa/ld200607/ldselect/ldsctech/166/16604.htm>

Hanksi I et al 2012 Environmental biodiversity, human microbiota and allergy are interrelated. *Proc Nat Acad Sci USA* 109, 8334–9

Ellwood P, Asher M I, García-Marcos L, Williams H, Keil U, Robertson C, Nagel G, the ISAAC Phase III Study Group 2013 Do fast foods cause asthma, rhinoconjunctivitis and eczema? Global findings from the International Study of Asthma and Allergies in Childhood (ISAAC) Phase Three. Thorax 68, 351–60 doi:10.1136/thoraxjnl-2012-202285

Food poisoning (pages 59–60)

<http://www.cdc.gov/outbreaknet/outbreaks.html>

<http://wwwnc.cdc.gov/eid/article/5/5/99-0502_article.htm>

<http://www.en.wikipedia.org/wiki/List_of_foodborne_illness_outbreaks_in_the_United_States>

Human bacteria and probiotics (pages 60–61)

Anon 2012 The human microbiome. *Economist* 18 August 2012, 62–4

Limdi J K, O'Neill C and McLaughlin J 2006 Do probiotics have a therapeutic role in gastroenterology? *World J Gastroenterol* 12, 5447–57

Organic food safety (pages 61–63)

Avery A 2006 *The Truth about Organic Foods.* Henderson Communications, Chesterfield MO

Heuer O E, Pedersen K, Andersen J S, Madsen M 2001 Prevalence and antimicrobial susceptibility of thermophilic *Campylobacter* in organic and conventional broiler flocks. *Lett Appl Microbiol* 33, 269–74

Chui S, Beilei G, Zheng J, Meng J 2005 Prevalence and antimicrobial resistance of *Campylobacter* spp and *Salmonella* serovars in organic chickens from Maryland retail stores. *Appl Env Microbiol* 71, 4108–11

The safety of genetically modified food (page 63)

Séralini G-E et al 2012 Long-term toxicity of a Roundup herbicide and a Roundup-tolerant genetically modified maize. *Food and Chemical Toxicology* 50, 4221–31. See also <http://www.slideshare.net/Revkin/translation-of-french-science-academies-critique-of-controversial-gm-corn-study> and <http://www.efsa.europa.eu/en/efsajournal/pub/2910.htm>

Chapter 4: You are what you eat

The Tolpuddle Martyrs (page 65)

<http://www.tolpuddlemartyrs.org.uk/index.php?page=before-the-arrest>

Boer War recruits (page 65)

<http://www.forces-war-records.co.uk/Boer-War-Casualties>

Rosenbaum S, Crowdy J P 1992 British Army Recruits: 100 years of heights and weights. *JR Army Med Corps* 138, 81–6 <http://www.gracesguide.co.uk/Joseph_Rank>

The history of nutrition and the discovery of vitamins (pages 66–71)

Gratzer W 2005. *Terrors of the Table: The curious history of nutrition.* Oxford University Press, Oxford

Diet and chronic disease (pages 71–76)

Boffetta P et al 2010 Fruit and vegetable intake and overall cancer risk in the European Prospective Investigation Into Cancer and Nutrition (EPIC). *J Natl Cancer Inst* 102, 529–37

Crowe F L et al 2011 Fruit and vegetable intake and mortality from ischaemic heart disease: results from the European Prospective Investigation Into Cancer and Nutrition (EPIC). *European Heart Journal* 32, 1235–43

Bender D A 2006. The antioxidant paradox. *The Biochemist* October 2006, 9–12

Lin J, Cook N R, Albert C, Zaharris E, Gazanio J M, Van Denburgh M, Buring J E, Manson J 2009 Vitamins C and E and beta carotene supplementation and cancer risk: a randomized controlled trial. *J Natl Cancer Inst* 101, 14–23 <http://infor.cancerresearchuk.org/healthyliving/dietandhealthyeating/howdoweknow/?view=Printer Friendly>

Health claims (pages 76–78)

Goldacre B 2008 *Bad Science*. Fourth Estate, London

Fish oils and the brain in children (pages 77–78)

<http://www.badscience.net/category/fish-oil/>

<http://www.apraxia-kids.org/atf/cf/%7B145BA46F-29A0-4D12-8214-8327DCBAF0A4%7D/richardson.pdf>

<http://www.dyslexic.com/articlecontent.asp?CAT=Dyslexia%20Information&slug%20=6936&title=Dispelling%20Dyslexia%20with%20Omega-3:%20Fishy%20or%20For%20Real?>

Epigenetics (pages 79–80)

Carey N 2011 *The Epigenetics Revolution*. Icon Books, London

Kaati G, Bygren L O, Edvinsson S 2002 Cardiovascular and diabetes mortality determined by nutrition during parents' and grandparents'

slow growth period. *Eur J Hum Genet* 10, 682–8.
<http://www.time.com/time/magazine/article/0,9171,1952313,00.html>

Obesity (pages 80–85)

Anon 2011 Urgently needed: a framework convention for obesity control. *Lancet* 378, 741. Editorial at the start of a special issue on obesity, with three 'comment' articles (pages 743–7) and four 'series' articles (pages 804–47)

Anon 2012 *The Big Picture Economist* 12 December 2012. A special report on obesity 1–14
<http://www.euro.who.int/ENHIS%20http://www.who.int/mediacentre/factsheets/fs311/en/>

Jebb S 2012 A system-wide challenge for UK food policy *BMJ* 344: e3414

Tax on unhealthy foods (page 86)

Mytton O T, Clarke D, Rayner M 2012 Taxing unhealthy food and drinks to improve health. *BMJ* 344: e2931

Marketing of food to children (page 86)

Hasting G 2012 *The Marketing Matrix*. Routledge, London and New York

Chapter 5: Feeding the nine billion

The future of food production and consumption (pages 87–89)

Godfray H C J et al 2010 Food security: the challenge of feeding 9 billion people. *Science* 327, 812–18

Anon 2011 The 9 billion-people question. *Economist* special report. 26 February 2011, 3–18 online at <http://www.economist.com/node/18200618>

Conway G 1997 *The Doubly Green Revolution: Food for all in the 21st century*. Penguin, London

Global net primary production (page 90)

Imhoff M L et al 2004 Global patterns in human consumption of net primary production. *Nature* 429, 870–3

Fertilizers (page 91)

Smil V 2001 *Enriching the Earth*. MIT Press, Cambridge MA

Energetics of food consumption (page 94)

May R M 1975 Energy cost of food gathering. *Nature* 255, 669

Malawi (pages 99–100)

Anon 2011 Malawi, fertilizer subsidies and the World Bank <http://www.web.worldbank.org/WBSITE/EXTERNAL/COUNTRIES/AFRICAEXT/MALAWIEXTN/0,contentMDK:21575335~pagePK:141137~piPK:141127~theSitePK:355870,00.html>

Patel R 2011 Hunger management. *New Statesman* 27 June 2011, 28–9

Anon 2008 Can it feed itself? *Economist* 1 May 2008

Juma C. 2011 *The New Harvest: Agricultural innovation in Africa.* Oxford University Press, New York NY

Collier P 2010 *The Plundered Planet* Oxford University Press, New York.

Land sparing and organic farming (pages 101–102)

Hole D G, Perkins A J, Wilson J D, Alexander I H, Grice F, Evans A D 2005 Does organic farming benefit biodiversity? *Biol Conservation* 122, 113–30

Phalan B, Onia M, Balmford A, Green R E 2011 Reconciling food production and biodiversity conservation: land sharing and land sparing compared. *Science* 333, 1289–91

Hodgson J A, Kunin W E, Thomas C D, Benton T G and Gabriel D 2010 Comparing organic farming and land sparing: optimizing yield and butterfly populations at a landscape scale. *Ecology Letters* 13, 1358–67

Use and benefits of GM

<http://www.isaaa.org/resources/publications/briefs/44/highlights/default.asp>

Carpenter J E 2010 Peer-reviewed surveys indicate positive impact of commercialized GM crops. *Nature Biotech* 28, 319–21

Rejection of GM maize as food aid by Zambia (page 106)

Paarlberg R 2008 *Starved for Science—how biotechnology is being kept out of Africa.* Harvard University Press, Cambridge MA

Fish stocks (page 106)

Pauly D, Christensen V, Guénette S, Pitcher T J, Sumaila U R, Walters C J, Watson R, Zeller D 2002 Towards sustainability in world fisheries. *Nature* 418, 689–95

Biofuels and food (page 108)

Searchinger T D et al 2009 Fixing a critical climate accounting error. *Science* 326, 527–8

Climate change and diet (pages 109–110)

Audsley E, Brander M, Chaterton J, Murphy-Bokern D, Webster C, Williams A 2009 *How low can we go? An assessment of greenhouse gas emissions from the UK food system and the scope for reduction by 2050*. WWF-UK

Godfray H C J, Pretty J, Thomas S M, Warham E J, Beddington J R 2011 Linking policy on climate and food. *Science* 331, 1013–14 <http://www.downloads.theccc.org.uk.s3.amazonaws.com/4th%20 Budget/4th-Budget_Chapter7.pdf> <http://www.theccc.org.uk/ pdf/TSO-ClimateChange.pdf>

Food

Further reading

The following are useful encyclopedic sources:

Harold McGee *McGee on Food and Cooking* (2004) Hodder and
    Stoughton, London
Kenneth F. Kiple and Kreimheld Conee Ornelas *The Cambridge World
    History of Food* (2000) Cambridge University Press, Cambridge
Alan Davidson *The Oxford Companion to Food* (1999) Oxford
    University Press

Books that give an interesting perspective on particular issues:

David A. Kessler *The End of Overeating* (2009) Penguin, London.
    A former head of the US Food and Drug Administration's view of
    the obesity crisis and the food industry
Jeffrey Steingarten *The man who ate everything* (1997) and *It must've
    been something I ate* (2002) both Alfred A. Knopf Inc., New York.
    A witty and informed food writer's collected essays, some of which
    are very memorable
Vaclav Smil *Enriching the Earth* (2001) MIT Press. The history of
    fertilizers and food production
Robert Paarlberg *Starved for Science—how biotechnology is being kept
    out of Africa* (2008) Harvard University Press. A social scientist's
    hard-hitting account of the unfulfilled potential of biotechnology
    in Africa
Calestous Juma *The New Harvest: Agricultural innovation in Africa*
    (2011) Oxford University Press. A distinguished Kenyan-born

academic gives an upbeat view of how Africa can increase its food production by harnessing modern technology

Hugh Pennington *When Food Kills* (2003) Oxford University Press. A very entertaining and idiosyncratic account of food hazards by an eminent microbiologist

Andrew Rimas and Evan D. G. Fraser *Empires of Food: Feast, Famine and the Rise and Fall of Civilizations* (2011) Arrow Books. History, geography, and economics of food

Marion Nestle *Food Politics: How the Food Industry Influences Nutrition and Health* (2007) University of California Press. An interesting polemic on the role of the food industry

Peter Gluckman and Mark Hanson *Mismatch: The lifestyle diseases timebomb* (2006) Oxford University Press. An account of how we are ill adapted to cope with modern environments, with disastrous consequences for our health

Stephen J. Simpson and David Raubenheimer *The Nature of Nutrition* (2012) Princeton University Press. An account of nutrition in the animal kingdom, focusing in part on the role of protein intake. The latter part of the book considers human obesity in this light

Larry Zuckerman *The potato: how the humble spud rescued the western world.* (1999) North Point Press New York. An entertaining and eclectic history of the potato as a staple food.

Useful websites include:

The UK Food Standards Agency <http://www.food.gov.uk/>
The World Health Organization <http://www.who.int/en/>
Cancer Research UK <http://www.cancerresearchuk.org/home/>

Index

A

Accum, F. 50
acrylamide 54
ADHD (attention-deficit hyperactivity disorder) 55
adulteration of food 50–1
Africa, food production 98–100, 106
agriculture
  effect of climate change 97
  evolution 11–15
  fertilizers for 91–2
  genetically modified crops 102–6
  green revolution 92–6
  livestock 110–11
  socio-political issues 100–1
  technology 99–100
aid to poor countries 101
allergies 56–9
Ames, B. 51–2
anaphylaxis 57
anatomy
  affected by farming 14
  change through cooking 9–11
ancestors of humans 2–4
  change in anatomy through cooking 9–11
  clues to diet 4–7
animal farming 110–11

increase 93
animals, domestication 13
antibodies 57
antioxidants 73–4, 77
Appert, N. (appertization) 17–18
aquaculture 107
aromas 24–6
artificial selection 12
associative learning 31, 32
Australopithecines, brain size 9
Australopithecus afarensis 3–4, 5, 6
aversion 32

B

babies, nutrition 79
bacteria
  food poisoning 59–60
  for fermentation 16
  in genetically modified crops 103
  in the human body 60–1
  killed by spicy food 38–9
  in organic food 62–3
balanced diet 67
bananas 25
Barker, D. 79
beef, BSE crisis 43–7
beer, safety 49–50
Bell, D. 92
Bernard, C. 67

Berthelot, C. L. 67
biodiversity 101–2
biomass fuel 108–9
biotechnology 102–6
Birdseye, C. 18
bitter taste, genetic variation
    29–30
blueberries 77
Blumenthal, H. 27–8
body mass index (BMI) 82–3
Borlaug, N. 92
brain function, fish oils for 78–9
brain size 9
brassicas 29
bread, origins 19
breastfeeding and food allergies 58
broad beans, intolerance 35–6
BSE (bovine spongiform
    encephalopathy) 43–7

C

cacao beans 19–20
calories 84
cancer prevention 73–4, 76, 77
canned food 17–18
cannibalism 45
carbon atoms 5–7
cardiovascular disease 75–6, 78
chemicals
    additives 50–1
    effect on children 55–6
    in food 51–3
chewing gum 24
children, obesity 82
chilli peppers 30–1
    heat in 23
    origin 40
Chinese Restaurant Syndrome
    23–4
chocolate, processing 19–20
citrus fruit, cure for scurvy
    68–9
climate, global 14–15
climate change 96–7

Climate Change Committee,
    UK 110
Collier, P. 100–1
colours 27
cooking, evolution 7–11
crisps, sensory experiment 27
crops 13
    height 92–3
cultural taboos 33–4

D

Darwin, C. 12
David, E. 40
Denisova fossil 3
Diamond, J. 13–14
diet
    environmentally friendly 110–11
    healthy 77–9
    regimes 85
disgust 31–2
distribution of food 88–9
DNA 80
    in fossil remains 2–3
    in genetically modified crops 103
domestication of plants and
    animals 13
Donkin, B. and J. Hall 18
Dorrell, S. 43
Durand, P. 18

E

E. coli outbreak in Germany
    51, 60
elements, early concept of 66–7
energy efficiency 94–5
English traditional cuisine 39–40
environmental impact
    of biomass technology 108–9
    of farming 94–5
    of genetically modified crops 105
    of livestock farming 110–11
epidemiological studies 74–6
epigenetics 79–80

evolution, human 2–3
  affected by cooking 9–11
exercise 85

F

farming *see* agriculture
fast food 84
fatty acids 77–8
favism 35–6
fermentation 16
fertilizers 91–2
  subsidies 99
Finland, food allergies 58
fire, evidence of use by early
    humans 9–10
fish 106–7
fish oils 77–8
flavour enhancers 23–4
flavours 24–6
  genetic variation 28–30
food poisoning 59–60
  elimination by cooking 8
fossil remains 2–4
  teeth 4–7
Fox, A. L. 29
freezing food 18–19
fruit, cure for scurvy 68–9
fuel, crops grown for 108–9
Funk, C. 70

G

Gajdusek, C. 45
Galen (Claudius Galenus) 66–7
genetic variation 28–30
  in intolerances 35–6
genetically modified (GM)
    food 62–3, 90, 102–6
genetics, epigenetics 79–80
Gentilcore, D. 40–1
Germany
  E. coli outbreak 51, 60
  food allergies 58
global climate 14–15

glutamate as a flavour
    enhancer 23–4
GM (genetically modified)
    food 62–3, 90, 102–6
golden rice 104
grandparents, nutrition 80
green revolution 92–6

H

Haber Bosch process 91–2
Harris, M. 33–4
health claims 77–9
hearing, sense of 27–8
heart health 75–6, 78
height, average 65–6
Hindus 33
Homo erectus 3
  brain size 9
Homo habilis 3, 5
  brain size 9
Homo heidelbergensis 3
Homo sapiens, brain size 9
human ancestors 2–4
  change in anatomy through
      cooking 9–11
  clues to diet 4–7
human evolution 2–3
  affected by cooking 9–11
human milk 31
humours of the body 66–7
hunter-gatherers 11–12, 14
hygiene hypothesis 58

I

Ikeda, K. 23
illness, association with food 32
imazalil 52
immune system, food allergies
    56–9
in vitro studies 73–4
international foods 21–2
intolerances 34–6
iron deficiency 71

isotopes of carbon 5–7
Italy, national cuisine 40–1

J

Jews 33
Juma, C. 99–100
junk food 84–6

K

Kessler, D. 84
kuru disease 45

L

lactase persistence 34–5, 36
Lawes, Sir J. B. 91
Leakey, L. 3
local cuisine 39–41
long-delay learning 32
Lyon Heart Study 75–6

M

Mad Cow disease 43–7
maize, yield gap 99
malaria 30, 35–6
Malawi, yield gap 99
malnutrition 65–6
    obesity 81–6
    in pregnancy 79
    vitamin deficiency 68–70
Malthus, T. 87–8
mammals, teeth 4
meat consumption 110–11
Mediterranean diet 75–6
milk
    human 31
    intolerance 34–5, 36
    powdered 50–51
    safety risks 48–9
minerals 71, 72
monosodium glutamate (MSG)
    23–4

mothers, nutrition 79–80
music as accompaniment 27
Muslims 33

N

Napoleon Bonaparte 17
national dishes 39–41
natural diet 1
natural selection 10, 12
natural solutions for increased food
    production 102
Neanderthals 3
nitrogen fertilizers 91–2
Nutcracker Man 6
nutrition 68–72, 77–9
    early concepts 66–8
    improving in Britain 65–6
    studies 73–6
nutrition value of cooked food 8
nutritional wisdom 81

O

obesity 81–6
odour receptors 25–6
oranges, pesticides on 52
organic farming 102
organic food 61–3, 77
Oxford, England 21–2
oysters 27–8

P

paleo diet 1
Paracelsus 52
Paranthropus boisei 6
Paranthropus robustus 6
Pasteur, L. (pasteurization) 8, 17
Pauly, D. 106
Pavlov, I. P. 32
pepper 36–9
pesticides 51–3
    in organic food 61–2
    resistance to 105

phenyl thiocarbamate (PTC) 29
photosynthesis 5–6
pickled food 16
plant growth 89–97
plant secondary compounds 8
plants, domestication 13
plasmids 103
Pliny the Elder 16
poisons, bitter taste 29–30
polymorphism 36
population increase 87–9
potatoes, toxins in 53
preferences in food 31
pregnancy
  cravings 81
  malnutrition in 79
preservation 15–19
processing 19–20
production of food, increasing
    89–107
prospective studies 74
protein content 84
Prusiner, S. 44

Q

quinine 30

R

raw food diet 7, 8
religious taboos 33
retronasal scent 25
retrospective studies 74, 75
rickets 72
Rozin, P. 30

S

safety 51–4
  in additives 50–1, 55–6
  in beef 43–7
  in beer 49–50
  cooking for 8
  in milk 48–9

in organic food 62
salt for preservation 16–17
saturated fats 78
savoury taste 23–4
Scoville Heat Units 30
scurvy 68
senses 27–8
  genetic variation 28–30
sensory-specific satiety 28
Shepherd, G. 25
Sherman, P. 38–9
sight, sense of 27
smell receptors 25–6
Smil, V. 92
Smollett, T. 48
socio-political issues 100–1
sounds 27–8
Southampton study 55–6
Spice Islands 37–8
spicy foods 23, 30–1, 36–9
starchy foods, acrylamide 54
Starling, A. and J. Stock 14
Steiner, R. 61
Streptococcus thermophilae 16
subsistence farming 99, 100
super foods 77
supertasters 29
Swift, J. 92
synaesthesia 26

T

taboos 33–4
tannins 29
taste, genetic variation 28–30
taste receptors 22–3, 26
tasters 29
teeth 4–5
  chemicals in enamel 5–7
  decay in farmers 14
ten per cent rule 109
toffee experiment 27
Tolpuddle Martyrs 65
tomatoes in Italian cuisine 40–1
tools, clue to ancestral diet 5

toxins
  bitter taste 29–30
  elimination by cooking 8
  size of dosage 52–4
Treaty of Tordesillas 38
TSEs (transmissible spongiform
    encephalopathies) 44–5

U

umami 23–4

V

Van Houten, C. 20
variety 28
vCJD (variant Creutzfeld Jacob
    Disease) 43–7
vegetables, bitter taste 29
vegetarian diet 110–11
viral food poisoning 59
viruses 44
visual input 27

vitamins
  vitamin A 70, 72, 104
  vitamin B 70
  vitamin C 52, 69, 70–1
  vitamin D 70, 71–2
  deficiency 68–70, 71–2
  discovery 70–1
von Liebig, J. 67–8, 91

W

wastage 107
water use in agriculture 94
weight gain 28
weight loss 85
wine tasting 27
World Health Organization 81, 83
Wrangham, R. 7, 9

Y

yeast 19
yield gap 98–9

Food